# essential atlas
## of astronomy

**BARRON'S**

Original title of the book in Spanish: *Atlas de Astronomía*
© Copyright 2001 by Parramón Ediciones, S.A., World Rights
Published by Parramón Ediciones, S.A., Barcelona, Spain

Authors: Parramón's Editorial Team
Illustrations: Parramón's Archives, Boreal, Jaume Farrés, Kobal, Josep Torres
Text: José Tola

First English-language edition for the United States, Canada, its territories and possessions published
in 2002 by Barron's Educational Series, Inc.
English-language edition © Copyright 2002 by Barron's Educational Series, Inc.

*All inquiries should be addressed to:*
Barron's Educational Series, Inc.
250 Wireless Boulevard
Hauppauge, New York 11788
**http://www.barronseduc.com**

International Standard Book Number 0-7641-2276-2

Library of Congress Catalog Card Number 2002100066

Printed in Spain
9 8 7 6 5 4 3 2 1

# FOREWORD

This *Essential Atlas of Astronomy* offers readers a wonderful opportunity to become familiar with the universe, its origin, its evolution, and the characteristics of the various heavenly bodies. It thus constitutes an extremely useful tool in approaching the marvels of the sky, with its mysteries, its apparently changeless laws, and the unending surprises that it offers.

The various sections of this book make up a complete compendium of astronomy. They include many illustrations and numerous concise but accurate charts that show the main characteristics of the celestial bodies, the history of their study, the instruments used for observation, and the fascinating adventure of their exploration and conquest. These illustrations, which are the core of this volume, are complemented by brief explanations and notes that aid in understanding the main concepts; there is also an alphabetical index that makes it easy to locate any topic of interest.

In undertaking this *Essential Atlas of Astronomy* we have set the goals of producing a practical and instructive book that is useful and accessible, characterized by strict scientific accuracy, and at the same time, enjoyable and clear. We hope the readers agree that we have achieved those goals.

# CONTENTS

# INTRODUCTION

## ASTRONOMY

Astronomy is what we call the science that is devoted to the **study of heavenly bodies**, plus all the general phenomena that take place outside our planet. Nowadays it is a science that makes use of complicated technology; it requires difficult mathematical calculations, and it involves travel in space, but its origins were very different.

The first astronomers lived about 5,000 years ago in Mesopotamia. They were **priests** who contemplated the sky and managed to predict eclipses. In addition, by determining the duration of the phases of the moon and the seasons—which had never before been measured, even though they were a permanent part of everyone's life—they improved the **agricultural practices** employed at that time. This seemingly small advance meant that the work in the fields could be linked to the cycle of the seasons and that it was possible to predict changes.

However, these astronomer-priests were entirely ignorant of the mechanisms that governed those celestial phenomena, and they interpreted them as **interventions by the gods.** In that way, astronomy was born in close association with religion and mythology.

The ancient **Greeks**, who created science as we recognize it today, also studied the sky and devised an explanation for the mysterious phenomenon of eclipses. In addition, with the aid of their calculations on heavenly bodies, they were able for the first time to determine the precise radius of the Earth.

Other people in ancient times, notably the Indians and the Egyptians, also devoted themselves to this

It is estimated that some fifteen billion years have passed since the formation of the universe with a huge explosion known as the Big Bang.

branch of knowledge, and for the same reasons. **Ptolemy** used his observations and measurements to establish an overall system in which the Earth was the absolute center and all the planets and other heavenly bodies rotated around it. This idea remained in force as dogma through nearly 1,500 years.

In medieval Europe, astronomy stagnated with no new developments; but in America the **Aztec** astronomers carried out minute observations of the heavens that allowed them to establish very precise **calendars** and perform mathematical calculations relative to the heavenly bodies. Around the sixteenth century the situation began to change. **Copernicus** was the first one to take in the new ideas from scientific thought that were being generated, and after twenty-five years of observations, he arrived at the conclusion that it was the Sun, rather than the Earth, that was the center of the universe. This was a completely revolutionary idea for the time, and it also marked the birth of modern astronomy.

## MODERN ASTRONOMY

At the time of Copernicus the heavens came under new scrutiny using **scientific criteria** and new instruments that were being invented. **Telescopes** finally made it possible to make scientific observations. Some great names from these times are Tycho Brahe, Kepler, Galileo, and Newton, who contributed enormously to the abandonment of the idea that the Earth was the center of the universe, thereby confirming the ideas of Copernicus.

As improvements were made in the instruments used for observing the skies, new celestial bodies were discovered, including the satellites of various planets, and it became possible to calculate the orbit of comets. It finally became possible to confirm with certainty the orbit of the planets that make up the **solar system**, and to begin studying other systems. In that way, astronomy situated our planet in a universe that kept expanding as more distant worlds were explored.

We will see what the overall solar system is like, and we will visit each of the **nine known planets** that make it up (including our own), as well as the

A comet is a heavenly body that has a very distinct appearance and behavior, and that travels through the solar system.

Moon and the asteroid belt. Spaceships have reached some of these celestial bodies and have taken samples of their soil; as a result, ever since the middle of the twentieth century we have known data that a mere century ago seemed absolutely impossible to obtain.

But our solar system is not the only one, and our star, the Sun, is just a medium-sized one located at the edge of our galaxy, which is one of many that make up the universe. Astronomy studies all these heavenly bodies and the phenomena that occur outside our small planet; these include comets, galaxies and nebulae, dwarf stars, supernovas, and the mysterious black holes.

The nine planets of the solar system, from left to right: Mercury, Venus, Earth, Mars, Jupiter, Saturn, Uranus, Neptune, and Pluto

## ASTRONOMICAL INSTRUMENTS

Professional astronomers use powerful **computers** to do their calculations, and complicated mathematical formulas to establish their theories; they also use enormous **telescopes** to observe outer space. But amateur astronomers can study the skies without such advanced methods, using anything from strong binoculars, which allow viewing the surface of the Moon in greater detail, to small telescopes that make it possible to view some distant galaxies. These instruments are all you need to enjoy this hobby.

In addition to the telescopes that are used for observing the light emitted by heavenly bodies and stars—in other words, optical instruments—the discovery of radio and other types of waves has opened up new fields of astronomy. Thus, many of the telescopes currently used in observatories use radio waves that are emitted from the farthest areas of the universe, and that take millions of years to reach us here on Earth. This branch of astronomy goes by the name of **radio astronomy**; it allows astronomers and scientists to reach more distant areas than those that are accessible using traditional optical telescopes.

In the section devoted to instruments used for observation we will also dedicate some space to the great **astronomers** who have made observations throughout the centuries and established theories that make it possible for us to have a more complete idea of what the universe is today.

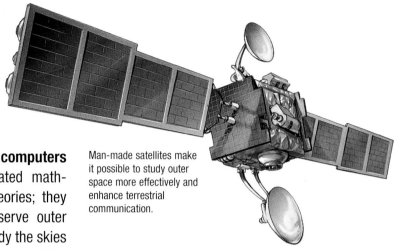

Man-made satellites make it possible to study outer space more effectively and enhance terrestrial communication.

## ASTRONAUTICS

Astronautics is a set of very diverse disciplines that are focused primarily on **space travel** beyond our planet. Even though this pursuit has a history, it's been around for scarcely half a century; nevertheless, it has added tremendously to our knowledge of astronomy and has contributed significantly to progress in many daily activities.

Some aspects of astronautics are as spectacular as putting a man on the Moon, or obtaining images of the surface of the planet Mars, but there are others as well that we scarcely realize, and that make our lives easier. These primarily involve **communications.** Nowadays, television reaches all parts of the globe thanks to the man-made satellites that rotate around the Earth. **Weather forecasting** is more accurate, and shows real maps with the location of weather fronts and their movement, thanks to the photos sent back by meteorological satellites. Cellular phones, which are so common today, allow us to speak with any point on the planet, thanks once again to these communications satellites. We will see what the first attempts to launch objects into space were like, with the successes and failures. But all of these attempts produced results that allowed another step forward. Just a little more than twenty years passed from the time of the first **man-made satellite** to the arrival of the first man to step onto the surface of the Moon; this is a very short period of time when we compare it to the centuries it took to gather the knowledge that made that voyage possible.

Astronautics offers a fascinating world of spaceships that are ever larger and more powerful, the shuttles that transport astronauts from a base on Earth to the **space stations** in orbit around the planet, and the space stations themselves, where the astronauts remain for months carrying out observations and experiments to make even longer journeys possible.

The Russian astronaut Valery V. Poliakov set the record for remaining in space for 437 days in 1993 and 1994.

## THE FUTURE OF ASTRONOMY AND ASTRONAUTICS

The speed at which advances in astronomy and astronautics take place makes it difficult to predict the future, but we will take a look at some outlines of the projects that are being prepared for the coming years. Bases on the Moon are already a possibility, and **permanent space stations** are a reality that we can read about in newspapers on a daily basis.

**Travel** to nearby planets is one of the projects underway, but exploration of the most distant ones and those beyond our solar system will be reserved for unmanned spaceships.

But scientific advances also produce a broad diversity of applications, so there are already some travel agencies that are beginning to organize trips to the Moon and stays on our satellite, where they are planning hotels and small cities. Although until a few years ago the majority of these plans were just **science fiction**, today they are starting to turn into **reality** (since the first space tourist took a flight in 2001); they will have repercussions in daily life for all of us because of the new uses that will be made of the technical resources needed to make them happen.

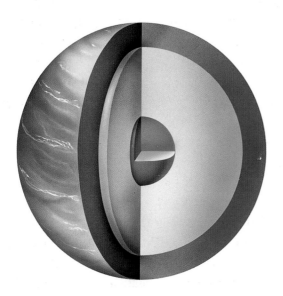

Planets are distinguished from stars by the fact that they generate no light of their own. The drawing shows the structure of Neptune, which has a rocky core, a middle layer of ice, and an outer layer of hydrogen and helium.

# THE ORIGIN OF EVERYTHING

Ever since the beginning of human evolution, people have been curious about the world that surrounds them. As their knowledge increased, these surroundings expanded to include the **heavens**. In an attempt to make sense of them, people devised scientific and religious theories. Each new explanation presented new problems. Today we are beginning to have a more accurate idea of how the **universe** originated.

The theory of the first explosion that formed the universe is known by the name of "**Big Bang**."

## THE GREAT EXPLOSION

The universe was a mass (1) of extremely dense, heavy matter—thousands of times heavier than the rocks that make up our planet. One day, about fifteen billion years ago, that mass of matter exploded (2) and began to send out fragments in all directions around it. The **galaxies** (3), **stars**, and other heavenly bodies were born from those fragments, and they are still moving farther apart from one another. The only things that are holding together are the units that make up a system such as galaxies, which contain thousands of stars. They all travel through space together and move farther away from other galaxies (4).

From the instant of the **initial explosion** the temperature of the universe has been decreasing; that is also true of the speed at which it is expanding.

# A PROBLEM AS OLD AS CIVILIZATION

When the ancients observed the starry sky at night and its slow movement, they wondered where those points of light were located and why they were moving. They thought that our planet was a kind of disk surrounded by spheres. The stars were attached to each of these spheres. The stars were on one, the planets on another; another was for the Sun, and yet another for the Moon. With the means available 5,000 years ago, this interpretation was accurate, and even though it strikes us as ridiculous today, it made it possible for us to come up with our present knowledge.

In order to find an answer to the mystery of how the universe was born, the ancients relied on "creators," which were supernatural gods in some religions, such as those of the Greeks and the Sumerians, or individual gods in other religions, as in Judaism, and later on, Christianity and Islam. In honor of these gods, they erected pyramids, altars, and monuments, from which the priests tried to communicate with the gods.

# SPACE EXPLORATION AND THE UNIVERSE

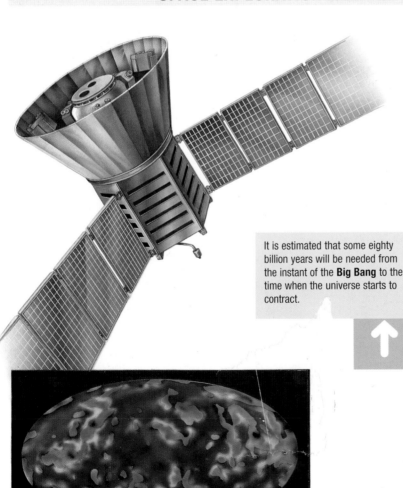

Three hundred thousand years after the Big Bang, the universe cooled off enough for the first atoms to be formed.

It is estimated that some eighty billion years will be needed from the instant of the **Big Bang** to the time when the universe starts to contract.

Thanks to **man-made satellites** and new **space stations**, it is possible to observe the universe without interference from the Earth's atmosphere. **Space telescopes** now in use are providing very valuable information that has made it possible to confirm the most recent theories. Thus, after that tremendous initial explosion, all original matter has been expanding and moving away, but there will come a time when that tendency will reverse and the universe will become more compact until it once again forms an infinitely dense mass. Then it will explode again and new stars and galaxies will be formed. This model is known as the "**Pulsating Universe**."

## TEMPERATURE

**Temperature** in the distant reaches of the universe is calculated by measuring the radiation that reaches us from there. The temperature is highest inside many stars and lowest in the spaces between them.

# THE LAWS OF THE UNIVERSE

The **Moon** revolves around the **Earth**, which revolves around the Sun. The **law of gravity** is responsible for this phenomenon and for the fact that the stars of the galaxies stay together and that comets travel through the entire universe along a trajectory that we can calculate. All heavenly bodies move in space according to the laws of **celestial mechanics**.

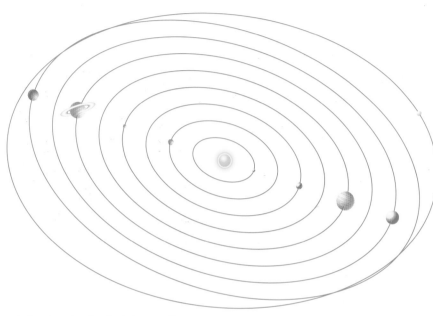

In the solar system, the planets trace an ellipse around the Sun.

## KEPLER'S LAWS

The ancients considered that the Earth was the center of the universe and that all the heavenly bodies revolved around it. **Copernicus**, in the sixteenth century, revolutionized astronomy when he declared that the Sun was the center and that the remaining planets, including the Earth, moved around it, although he could not explain how they did so. In the seventeenth century, it was **Kepler** who figured out how that movement worked. After many observations and calculations, he declared that the planets revolve around the Sun along **elliptical orbits**, and he explained how that works in three laws that bear his name.

### FIRST LAW

Each planet moves in an elliptical orbit and the Sun is one of the foci of the ellipse.

### SECOND LAW

The closer the planets are to the Sun in their trajectory, the faster they move.

### THIRD LAW

The further a planet is from the Sun, the greater its speed of revolution.

# THE LAW OF GRAVITY

In the seventeenth century, Newton explained why the planets revolve around the Sun along elliptical orbits. It was also **Newton** who said that when an apple falls to the ground it is because there is a force (**gravity**) inside the Earth that exerts an attraction on all objects. In addition, he affirmed that this phenomenon affects not only our planet, but also all the rest of the heavenly bodies. As a result, the Sun attracts the Earth, and the Earth attracts the Sun (but with much less force). If our planet doesn't get sucked into the Sun it is because there is another **force** (due to its movement) that counteracts it. The series of points at which both forces equalize one another makes up the **trajectory** (the **Earth's orbit**) in which our planet moves around the Sun. The same thing happens with the rest of the planets.

The formula for the **Universal Law of Gravitation** is:

$$F = G \, \frac{m \, m'}{r^2}$$

where F is the force between two bodies, m and m' are the masses of these two bodies, r is the distance that separates them, and G is a constant value (the constant of gravitation).

**Galaxies**, **nebulae**, and **planetary systems** remain stable thanks to **gravitational force**.

According to the **Universal Law of Gravitation**, two bodies attract one another in a way that is directly proportional to their mass and inversely proportional to the square of their distances.

# COMETS

**Comets** are one of the most stunning phenomena in space. Some of them appear at regular intervals, but others only after many centuries or once in their lifetime. These are traveling bodies that wander through space; when they come close to the **Sun** they commonly leave a large trail of light behind them that is known as a tail.

## WANDERING ICE BOULDERS

**Comets**, as we see them with the naked eye from the Earth, consist of a core surrounded by a bright area known as the head, plus frequently a long tail. The **core** is a mass of rocky fragments held together by ice that measures an average of 6 miles (10 km) in diameter, although there are some that are much larger. The **head** is made up of gas and dust, and it occupies an area far larger than the nucleus. Finally, the **tail** is the gas and dust from the head that trail behind the comet when **solar wind** acts on it in proximity to the sun. Not all comets have a tail, since if they are very small the amount of available gas and dust is not sufficient.

The **tail** of a comet always points away from the **Sun**, and even though it reaches its greatest length as it passes close to the Sun, that is when it is hardest to see from Earth.

### A LONG TAIL

The **tail** of a comet can extend for hundreds of millions of miles/kilometers behind the **head**. The longest known tail was observed in 1843; it filled half the sky.

The ice in comets is made up principally of methane, ammonia, carbon dioxide, and water.

When a comet (1) approaches the Sun, even though it moves in a direction around the Sun (3), the tail (4) always points away from the location of the Sun (2).

# METEORITE SHOWERS

When the **tail** of a **comet** comes close to the **Earth**, a very striking phenomenon is commonly produced, known as a **meteorite shower**. A great number of **falling** stars—far greater than normal—enter the night sky. The particles of rock and ice from the tail are vaporized as they come into contact with the **atmosphere**, and before disappearing they leave a trail of light, which we see as a shooting star. Any **meteorite** that comes into contact with the atmosphere turns into a shooting star when it disintegrates. Some showers of shooting stars occur at fixed times due to the fact that the Earth crosses the former path of a comet.

In 1993 the comet **Shoemacker-Levy 9** passed close to Jupiter, and the gravity pull from the planet broke it up into twenty pieces.

**Halley's Comet**, which came close to the Earth in 1986, experienced a collision with some celestial body in 1993 that made it increase in size. When it next approaches the Earth, in 2062, we will be able to see the effects of the collision.

On December 27, 1872 there occurred the greatest shower of falling stars known, when the Earth passed through the tail of the broken-up comet **Biela**.

## THE PRINCIPAL SHOOTING STARS

| Name | Frequency | Date* |
|---|---|---|
| Quadrantids | 3 per minute | January 3 |
| Aquarids | 2 per minute | May 4 |
| Perseids | 5 per minute | August 12 |
| Geminids | 5 per minute | December 13 |

*Day on which they are visible

# STARS: what they are like and how they evolve

Stars are masses of **incandescent gas** that are spread fairly evenly throughout space; some of them form groups that we can see in the night sky in the shape of small points of light. Some are brighter than others, but that is merely their appearance, since it depends on the distance at which the stars are located. Stars don't always remain the same; rather, they are born, they grow, and they die. Some, such as the Sun, have **planets** that revolve around them.

## BRIGHTNESS AND SIZE

When we look at the stars during the night, some appear brighter than others, but that has only to do with their appearance. The **brightness** that we observe depends on the **size** of the stars, their inherent brightness, and the **how far** they are from us. Thus, a very large, bright star that is far away looks much weaker to us than some other small one that is not very bright, but that is closer to us. As a result, every star has an **apparent size** (the brightness that we observe) and an **absolute size** (its true size).

On a moonless night and away from cities and other population centers, about 3,000 stars are visible to the naked eye. Using a small **telescope** you can see up to 300,000 stars.

## THE COLOR OF STARS

If we look closely, we see that not all stars are the same color.

Formerly, stars were classified in four colors: red, orange, yellow, and white. Each of those colors corresponds to the **temperature** in the star. The hottest stars are the white ones, and the least hot are red. This is similar to what happens to a piece of steel heated in a fire: first it becomes red, and as the temperature increases, the steel changes color until it becomes a bluish white. Modern **astronomers** distinguish seven main types of stars according to their temperature.

### TYPES OF STARS BASED ON THEIR COLOR

| Type | Color | Temperature ($^0$F) |
|---|---|---|
| O | blue-purple | 50,000–90,000 |
| B | blue-white | 18,000–50,000 |
| A | white | 13,500–18,000 |
| F | white-yellow | 10,800–13,300 |
| G | yellow | 9,000–10,800 |
| K | orange | 6,300–9,000 |
| M | red | 4,500–6,300 |

A **professional telescope** that can detect stars of magnitude 22 can pick up as many as three billion stars.

# THE BIRTH OF A STAR

Space is full of tiny particles of dispersed **atoms and matter**. This is referred to as spatial dust. In some places there are only three atoms per cubic meter, but in others there is enough material present for it to gradually condense around a point. Stars are born when spatial dust clumps together around a single point. When the mass reaches a certain size, the star begins to heat up inside, where the temperature can reach several million degrees. At that point it begins to give off light, and that is when we consider that a star has been born.

## CONTRACTIONS

When stars are young, they experience irregular contractions and give off a large quantity of particles, similar to what happens with **solar wind**.

If a young star keeps up its irregular contractions for very long, it uses up all its fuel and is extinguished after just a few million years.

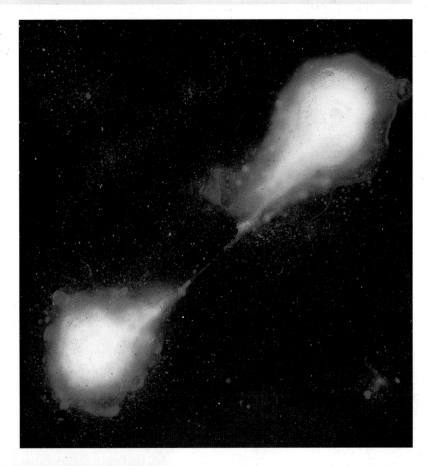

## GROWTH AND DEATH

The nucleus of a star is made up primarily of hydrogen, which is the fuel that keeps it active. When all the **hydrogen** has been used up, the star begins to degenerate. It starts to contract onto itself, and the bonds among the atoms are broken; the star then is like an "electron soup" that contains the nuclei of the atoms. At that point the star gives off a lot of light, but it begins to cool down. In this phase it uses helium (which is present in much smaller quantities) for fuel. The last phase is also "explosive," with a new emission of light before the star disintegrates into a cloud of interstellar matter, as if it were the smoke from some explosion.

Small stars are not as hot as large ones, so they "burn" more slowly and last longer.

# STARS: from element factories to black holes

There are a number of very important phenomena for the whole universe that take place inside stars. This involves the manufacture of the **chemical elements** that make up matter—in other words, **nuclear fusion**. Stars are also the origin of one of the most astonishing and mysterious phenomena in the universe: the famous **black holes**.

## THE CHEMICAL ELEMENTS

All planets, rocks, the air, and living creatures are made up of **chemical elements**. Some appear by themselves, such as **oxygen**, which is part of the atmosphere that we breathe (there are two oxygen atoms linked together); but many others join together to make up chemical compounds, such as **water** (which is made up of two hydrogen atoms and one of oxygen). **Hydrogen** is the simplest element of all; then comes **helium**. These two elements are the most abundant ones in the whole universe. They are the first ones that were formed. All the other ones were formed inside the stars, which therefore function like factories that produce the chemical elements.

The **helium** atom consists of a nucleus made up of two **protons** and two **neutrons**, plus two **electrons** that spin in an orbit around them.

The **hydrogen** atom consists of just one **proton** in the center and one **electron** that spins in an orbit around it.

**Carbon** has six **protons** and six **neutrons** in its nucleus. This elements was formed (and is still being formed) inside stars.

## THE COMBUSTION OF STARS

When we look at the sky we see the stars as tiny points that give off **light**. Light is **energy** that is produced inside the star through a process referred to as **nuclear fusion**. This involves a fusion between two or more **atoms** to produce a new atom whose **mass** is a little less than the sum of all the atoms that go into its formation. As a result, there is a little bit of leftover material that is transformed into energy; when that energy escapes the stars, it does so in the form of the light that we can see from the Earth.

### NUCLEAR FUSION

Scientists are trying to achieve **nuclear fusion** here on Earth to create energy without the hazardous wastes that are produced by the **nuclear power plants** currently in use.

The fusion of four **protons** produces one atom of **helium** plus a small amount of energy.

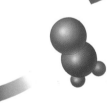

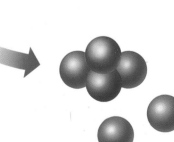

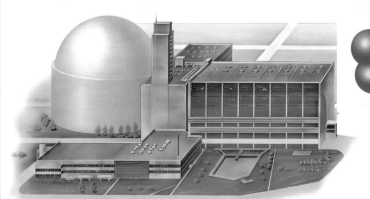

# BLACK HOLES

In very distant reaches of the universe astronomers have observed certain areas where their telescopes register no image. However, the calculations they have done indicate that there must be something there. Since no image appears in the photographs, scientists called these areas **black holes**. They are mysterious places, and they have therefore been studied intensely; it has been discovered that these involve **proton stars** of a density that is so great that their gravity keeps any kind of energy, even **light**, from leaving.

So far it has not been possible to observe a black hole directly; only its effects are visible.

The **mass** of a black hole is several times greater than that of the Sun, but a black hole is only a few miles/kilometers in diameter.

The force required to escape a black hole is greater than the **speed of light**.

# TYPES OF STARS

Although in theory all stars are equal, they appear different depending on their age, their size, and their evolution; therefore, they can be classified into several types. Amateur astronomers can also observe many of them using small telescopes. The main types are **double stars**, **variable stars**, **novas**, **supernovas**, **pulsars**, and **quasars**.

## DOUBLE STARS

In many parts of space there are stars that rotate together around a common center of gravity. These are known as **double stars**. These stars have a common origin in a mass of spatial matter that condensed and formed the two stars.

After a supernova explodes, there is some matter left over that forms clouds that are scattered through space.

## VARIABLE STARS

There are stars that don't always shine with the same brightness, but rather change at regular intervals that vary from as little as a few months to as long as several years. These are known as **variable stars**; they are distinguished from others that vary in brightness on an irregular basis because they are undergoing some type of change. Other stars vary in brightness because they are part of a pair of stars that rotate around one another; seen from the earth, they regularly eclipse one another.

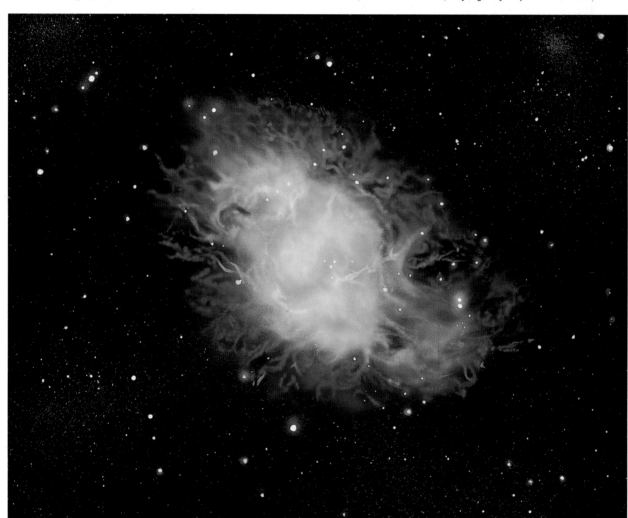

## NOVAS

In star pairs formed by a **red dwarf** and a **white giant**, sometimes the great force of attraction of the dwarf star draws hydrogen from the giant, and this additional fuel makes it shine very brightly for several hours. The red dwarf suddenly increases in brightness and is then known as a **nova**.

## SUPERNOVAS

In their last stages of **life**, stars are red in color. Giant stars explode spectacularly and increase in brightness several thousand times. This explosion is due to the fact that the **nuclear reactions** inside the star have completely used up the **hydrogen** and new, heavier elements have been produced. The mass is so great that the star implodes on itself and explodes, blasting all its matter through space.

Introduction

**Space**

The solar system

The Sun

Mercury

Venus

Earth

Mars

Asteroids

Jupiter

Saturn

Uranus

Neptune and Pluto

Exploring the universe

Astronautics

Alphabetical index

# PULSARS

**Pulsars** are **neutron stars** that are produced at the end of the life of a **giant star**, after it explodes. They rotate at high velocity (up to 600 times a second), and their **magnetic field** produces very strong electromagnetic currents. These waves reach Earth in the form of impulses that repeat at regular intervals, as if the star had a pulse, hence the name.

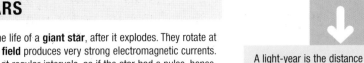

A light-year is the distance that light travels in one year at 183,000 miles/sec.

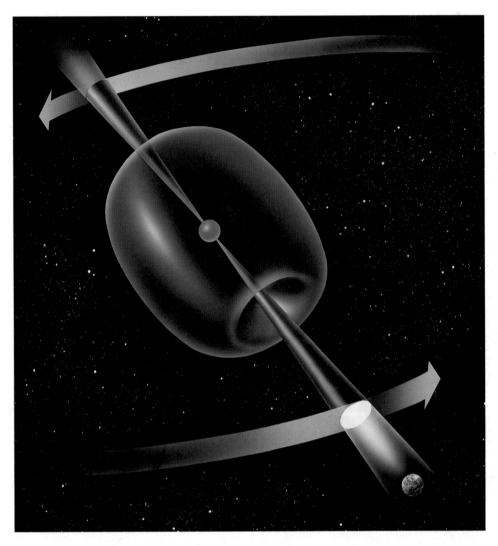

Pulsars are surrounded by a very strong magnetic field.

## QUASARS

Quasars were first discovered in the 1960s; they are very distant sources of electromagnetic radiation. It is believed that they are the most distant objects, and that they are moving at speeds of up to around 153,000 miles per second (250,000 km/sec). They could be the nuclei of new **galaxies** being formed, or the center of a **black hole**.

## BRIGHTEST STARS OF THE NORTHERN HEMISPHERE

| Name | Magnitude | Constellation | Distance (light-years) |
|------|-----------|---------------|------------------------|
| Arturus | −0.1 | Boötes | 36 |
| Vega | 0.0 | Lira | 27 |
| Cappella | 0.0 | Auriga | 45 |
| Procion | 0.4 | Canis Minor | 11 |
| Betelgeuse | variable | Orion | 520 |

## BRIGHTEST STARS OF THE SOUTHERN HEMISPHERE

| Name | Magnitude | Constellation | Distance (light-years) |
|------|-----------|---------------|------------------------|
| Sirius | −1.4 | Canis Major | 8.7 |
| Canopus | −0.7 | Carina | 650 |
| Alpha Centauri | −0.3 | Centaur | 4.3 |
| Rigel | 0.1 | Orion | 900 |
| Achernar | 0.5 | Eridanus | 118 |

# STAR CLUSTERS AND NEBULAE

Space is filled with matter that is distributed irregularly. The planets and stars originated from this interstellar matter, which congregates to form nebulae. After stars are formed, they almost never appear isolated, but rather in clusters of varying density. From the Earth, this **interstellar matter** and the groups of stars appear like disperse, colored clouds.

## STAR CLUSTERS

Stars don't appear in isolation, but rather in groups known as **clusters**. This is due to the fact that they were formed from the same mass of **interstellar matter**, which gave rise to several of them when it condensed around different points. Almost all the stars that make up a single cluster are of about the same age, and they move through space at the same speed. There are two types: Those that are formed by scattered stars and that are referred to as **open clusters**, and those that are made up by thousands of stars grouped very closely together in a type of sphere; these are known as **globular clusters**.

**Globular clusters** are found at a distance between 20,000 and 100,000 light-years from Earth and are composed of very old stars.

One of the most famous cumuli is the **Pleiades** in the constellation **Taurus**.

**Open clusters** are found at about 5,000 light-years from Earth; they are comprised of relatively young stars.

### SOME STAR CLUSTERS THAT CAN BE EASILY OBSERVED

| Constellation | Name | Type | Hemisphere | Observation |
|---|---|---|---|---|
| Ara | NGC6193 | open | Southern | with binoculars |
| Auriga | M38 | open | Northern | with binoculars |
| Cancer | M44 | open | Northern | with naked eye |
| Canis Major | M41 | open | Southern | with naked eye |
| Centaurus | NGC5139 | globular | Southern | with binoculars |
| Scorpio | M7 | open | Southern | with naked eye |
| Pegasus | M15 | globular | Northern | with naked eye |
| Taurus | M45 (Pléyades) | open | Northern | with naked eye |
| Vela | NGC2547 | open | Southern | with naked eye |

The Trifid nebula, in the constellation Sagittarius.

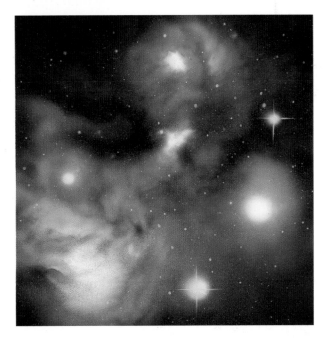
A bright nebula in the constellation Orion.

## NEBULAE

These structures are clouds of **gas** and **interstellar dust** that may or may not be visible from Earth, depending on their density. Some give off light when a nearby star warms them, but others have only dark gases, so they are not visible. Just the same, since interstellar dust absorbs light, we can deduce where a nebula is when it hides some other body in space whose existence can be confirmed by other means.

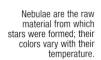
Nebulae are the raw material from which stars were formed; their colors vary with their temperature.

Nebulae that are made up of a cloud of gas surrounding a star are known as planetary nebulae.

### SOME NEBULAE THAT CAN BE EASILY OBSERVED

| Constellation | Name | Type | Hemisphere | Observation |
|---|---|---|---|---|
| Aquarius | NGC7293 | planetary | Southern | with binoculars |
| Carina | NGC3372 | dust | Southern | with binoculars |
| Lira | M57 | planetary | Northern | simple telescope |
| Orion | M42 | dust | Southern | with naked eye |
| Sagittarius | M20 | dust | Southern | simple telescope |
| Vulpecula | M27 | planetary | Northern | simple telescope |

# GALAXIES

The universe has been expanding ever since the occurrence of the **original great explosion**. The matter produced by this explosion is what has given rise to the **interstellar clouds** of gas and dust, the **stars**, and the **planets**. Yet all this matter and these celestial bodies are not distributed evenly throughout space, but are clustered together in groups. One of the most important groups is the **galaxy**, which is a basic solar system.

## TYPES OF GALAXIES

The tremendous number of stars that make up a galaxy can be classified in very different ways, and this has given rise to different types. Some are **round** or somewhat **elliptical**; others are flat or lens-shaped, and others are **spiral** in shape, with two or more arms that branch off from a central nucleus, which is where most of the stars that comprise it are located.

Galaxies, like star clusters and nebulae, are named using a letter and a number. The first classification was made by the French astronomer **C. Messier**, and that's why many galaxies are designated with an M and a number.

## THE COLLISION OF THE GALAXIES

Since many galaxies—mainly those that make up a **local group**—are located relatively close to one another, there are frequent "collisions." What really happens is that as they draw nearer to one another, the **gravitational pull** of each galaxy starts to influence the most distant stars of the other. That way, each galaxy loses a few of its stars, which are drawn toward the other galaxy. In the most extreme cases, both masses of stars end up forming a new galaxy.

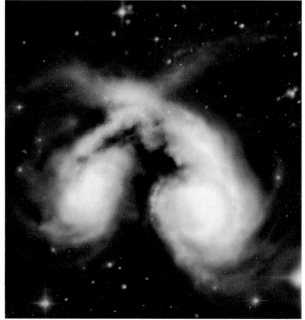

When two galaxies combine, some stars may collide with others, but it doesn't always happen that way.

## LOCAL GROUPS

It is not usual for galaxies to be alone in the universe; rather, they combine in units of two or more galaxies—these are known as **local groups**.

# EXPANSION OF THE UNIVERSE

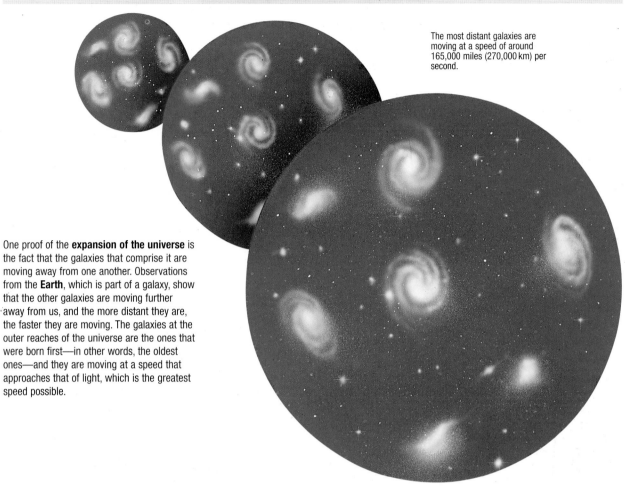

The most distant galaxies are moving at a speed of around 165,000 miles (270,000 km) per second.

One proof of the **expansion of the universe** is the fact that the galaxies that comprise it are moving away from one another. Observations from the **Earth**, which is part of a galaxy, show that the other galaxies are moving further away from us, and the more distant they are, the faster they are moving. The galaxies at the outer reaches of the universe are the ones that were born first—in other words, the oldest ones—and they are moving at a speed that approaches that of light, which is the greatest speed possible.

## HOW DO YOU MEASURE A GALAXY'S SPEED?

In order to measure the speed of a star or a galaxy, scientists use what's known as the **Doppler Effect**; when a heavenly body moves away from us, the light it gives off becomes more red the faster it goes, and conversely, when the object approaches us, the light it gives off is blue. This is the same effect that is produced on Earth with **sound**. The whistle of an approaching locomotive gets sharper as it comes closer, but once it passes us, the sound gets progressively deeper.

The galaxy Andromeda is 2.2 million light-years from Earth.

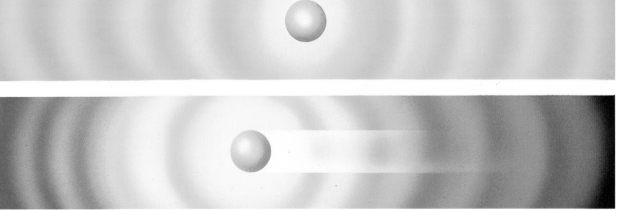

# NORTHERN HEMISPHERE CONSTELLATIONS

The stars that are visible on a clear night make up certain figures that are known as **constellations**. In ancient times astronomers observed this phenomenon and gave mythological names to these figures. They are still used today, and they are a helpful reference on **star maps**. In the Northern Hemisphere the stars are grouped into thirty-seven constellations.

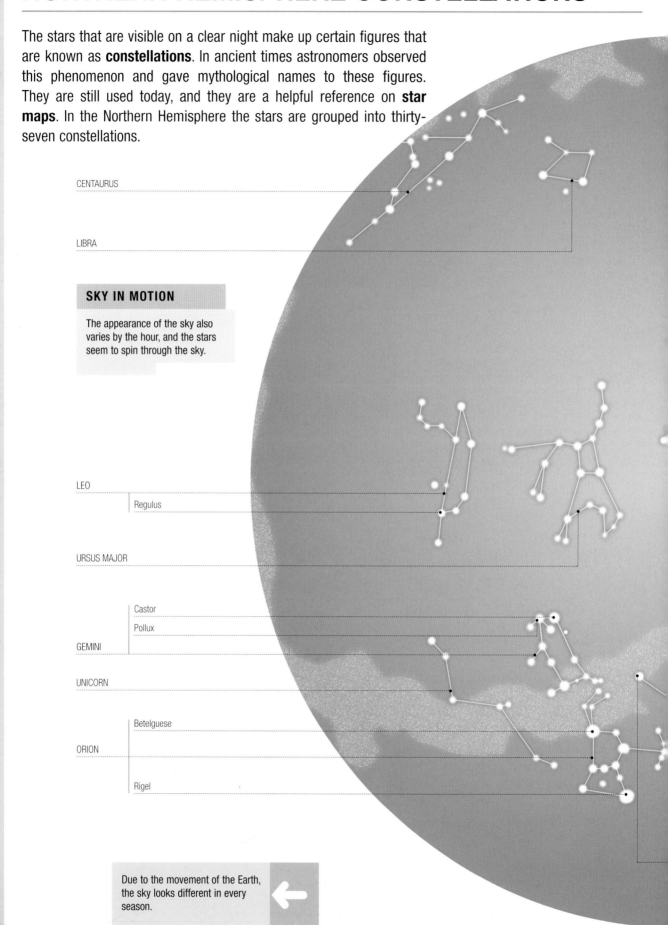

CENTAURUS

LIBRA

## SKY IN MOTION

The appearance of the sky also varies by the hour, and the stars seem to spin through the sky.

LEO

Regulus

URSUS MAJOR

Castor

Pollux

GEMINI

UNICORN

Betelguese

ORION

Rigel

Due to the movement of the Earth, the sky looks different in every season.

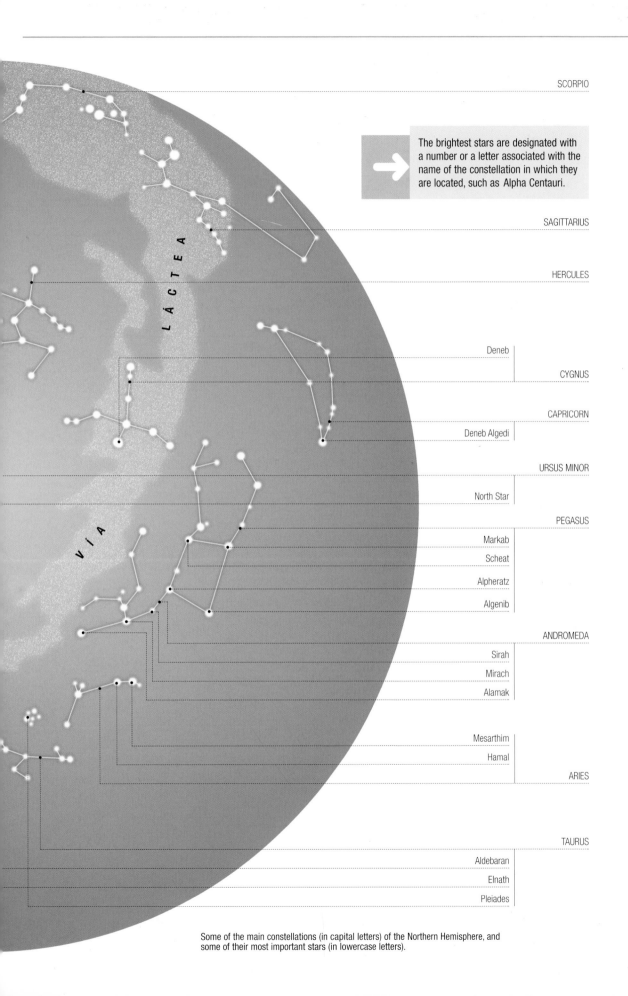

SCORPIO

> The brightest stars are designated with a number or a letter associated with the name of the constellation in which they are located, such as Alpha Centauri.

SAGITTARIUS

HERCULES

Deneb

CYGNUS

CAPRICORN

Deneb Algedi

URSUS MINOR

North Star

PEGASUS

Markab

Scheat

Alpheratz

Algenib

ANDROMEDA

Sirah

Mirach

Alamak

Mesarthim

Hamal

ARIES

TAURUS

Aldebaran

Elnath

Pleiades

Some of the main constellations (in capital letters) of the Northern Hemisphere, and some of their most important stars (in lowercase letters).

# SOUTHERN HEMISPHERE CONSTELLATIONS

Some of the fifty-one **constellations** of the Southern Hemisphere were known in ancient times and were also given mythological names, but most of them were discovered more recently; as a result, they often have names that pertain to scientific instruments, such as Telescope, Compass, Sextant, and so on.

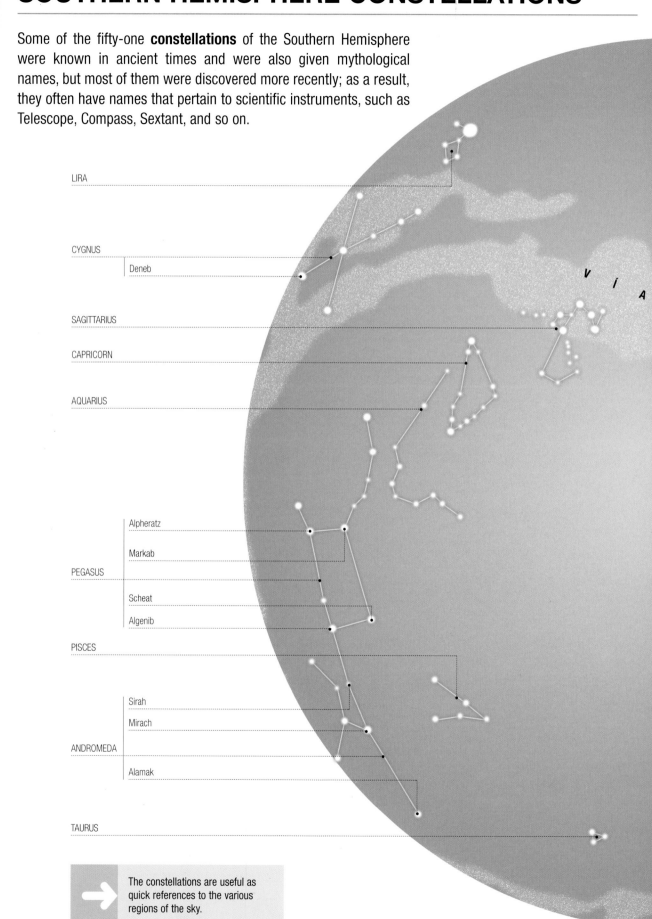

LIRA

CYGNUS

Deneb

SAGITTARIUS

CAPRICORN

AQUARIUS

Alpheratz

Markab

PEGASUS

Scheat

Algenib

PISCES

Sirah

Mirach

ANDROMEDA

Alamak

TAURUS

V i A

The constellations are useful as quick references to the various regions of the sky.

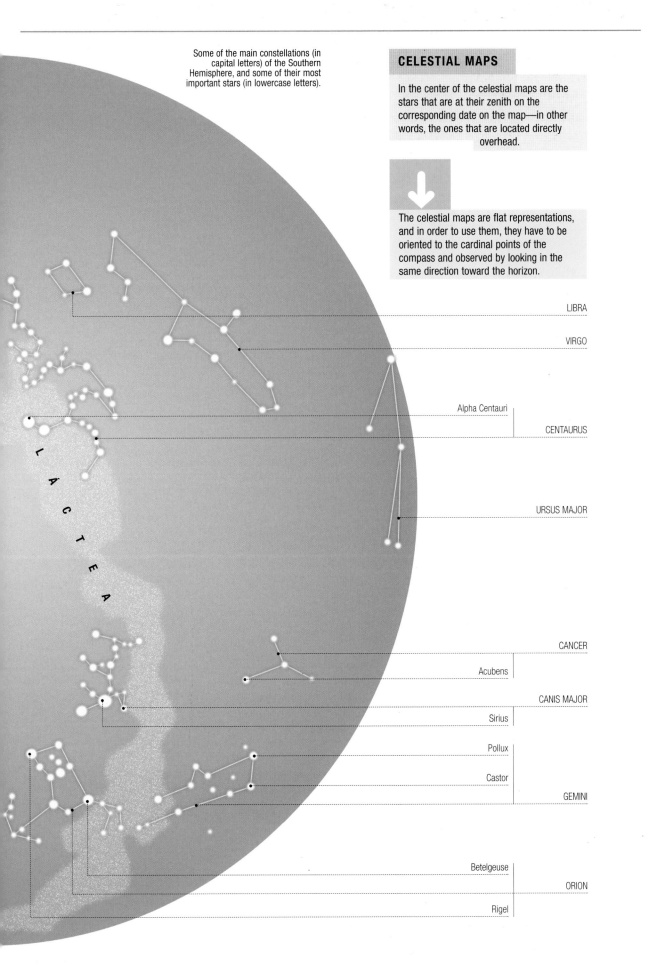

Some of the main constellations (in capital letters) of the Southern Hemisphere, and some of their most important stars (in lowercase letters).

## CELESTIAL MAPS

In the center of the celestial maps are the stars that are at their zenith on the corresponding date on the map—in other words, the ones that are located directly overhead.

The celestial maps are flat representations, and in order to use them, they have to be oriented to the cardinal points of the compass and observed by looking in the same direction toward the horizon.

LIBRA

VIRGO

Alpha Centauri

CENTAURUS

URSUS MAJOR

CANCER

Acubens

CANIS MAJOR

Sirius

Pollux

Castor

GEMINI

Betelgeuse

ORION

Rigel

L Á C T E A

# THE MILKY WAY

Among the many **galaxies** that make up the universe, the one we know best is the **Milky Way**, since that is where we are located. Since it forms a type of plane at one end, where the **Sun** and its planets are located, we see it in profile in the night sky, and it looks like a bright stripe that goes across the sky from one end to the other.

Image of the Milky Way obtained through stellar radiation.

## HISTORY

Aristotle thought that this whitish band that crosses the sky was an atmospheric disturbance. In the seventeenth century, when the first telescopes were constructed, Galileo discovered that this band is actually made up of stars, and is therefore not a phenomenon related to our atmosphere. Little by little, new stars were discovered, and from that time forward the Milky Way has become synonymous with the universe. But the arrival of the twentieth century produced a new scientific revolution. It was discovered that there are other enormous, distant clusters of stars—new galaxies—so the universe is not simply the Milky Way, but rather a large collection of galaxies, including ours.

## THE SHAPE OF OUR GALAXY

When we look at the Milky Way, it appears like a long stripe, but that is due merely to an optical effect, because we are on the same plane. The Milky Way is a spiral galaxy with a central area in the shape of a lens, or a flattened disk, and four arms that branch off from it. The Sun and the Earth are located at the outer edge of one of these arms.

Side view of the Milky Way.

The shape and the dimensions of the Milky Way have been determined by using both optical observations and measurements of radiation emitted from various points.

# EVOLUTION OF THE MILKY WAY

It appears that the Milky Way was originally a spherical galaxy that was spinning at low speed, then the interstellar matter of the core became more concentrated, increasing in density and speed of rotation. This caused a gradual flattening out to produce the appearance of a flattened disk that it exhibits today. As the stars evolved at the outer edges, the arms were formed until the entire assembly took on the shape of a spiral with four arms. This whole process evidently occurred in the course of about ten billion years.

It is calculated that the Milky Way contains around 300,000 stars. All the stars that we can see with the naked eye are part of our galaxy.

### DIMENSIONS

The central nucleus of the Milky Way has a diameter of around 15,000 light-years; the thickness of the central nucleus is around 2,000 light-years.

There are many galaxies in the universe (above right); our solar system is located at one end (lower left) of the Milky Way.

There are many clusters of stars outside the plane of the galaxy.

It appears that there is a large black hole in the center of our galaxy.

# ECLIPSES

In ancient times, eclipses often were omens of evil occurrences, for it was supposed that the disappearance of the light was a punishment from the gods. Today we understand that they are a phenomenon that is a product of the celestial workings, when two or more heavenly bodies line up in certain positions in their orbits, thereby interfering with the light that carries from one to another.

## ECLIPSES

Every heavenly body that revolves around a star always has one lighted side and one dark one; in addition, it projects behind it a broad cone where there is no light. This is what happens with our planet when it revolves around the **Sun**. At **night**, even if we traveled many miles/kilometers above the surface of the Earth, we would still be in the dark, since the projected cone of darkness is very long. When the **Moon** passes behind this cone, it too is obscured, and the result is a **lunar eclipse**. The converse occurs when our Moon comes between us and the Sun. In that case the result is a **solar eclipse**, but since the Moon is smaller, so too is its projected cone of darkness, and it doesn't cover the entire surface of the Earth, but only a certain circle.

### SOME LUNAR ECLIPSES IN THE DECADE FOLLOWING 2000

| Date | Type |
|------|------|
| January 9, 2001 | total |
| May 16, 2003 | total |
| November 9, 2003 | total |
| October 28, 2004 | total |
| March 3, 2007 | total |
| February 21, 2008 | total |
| August 16, 2008 | total |

Three instances in which the shadow cone of the Earth totally or partially obscures the Moon, our satellite.

In order for an eclipse to occur, the heavenly bodies in question have to be perfectly lined up.

The orbit of the Moon around the Earth is inclined about five degrees with respect to the orbit of the Earth around the Sun. This means that when the Moon is full, it passes above or below the shadow cone produced by the Earth. There are only a few times when the Moon is lined up exactly with the Earth and the Sun at a full moon, thereby producing an eclipse.

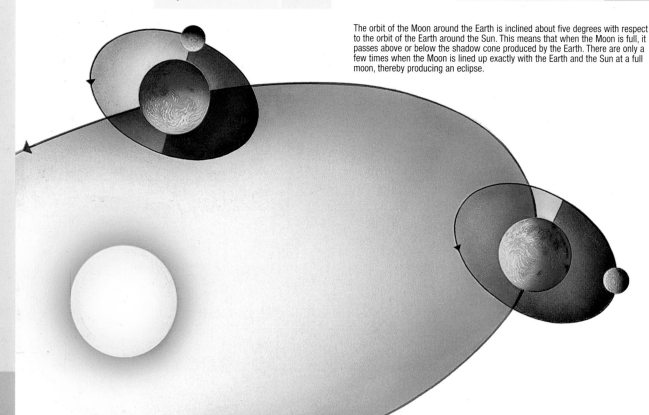

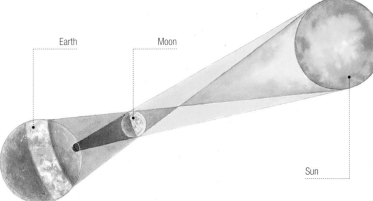

Earth

Moon

Sun

## SOLAR ECLIPSES

The occultation of solar light is one of the most spectacular astronomical phenomena that we can observe from Earth. When the Moon comes between the Sun and the Earth, it projects a circular shadow that turns day to night for a few minutes in the area of a **total eclipse**. In other areas, the Moon covers only a part of the solar disk, and the result is a **partial eclipse**. When the moon appears slightly smaller than the solar disk, it produces an **annular eclipse**.

Introduction

**Space**

The solar system

The Sun

Mercury

Venus

Earth

Mars

Asteroids

Jupiter

Saturn

Uranus

Neptune and Pluto

Exploring the universe

Astronautics

Alphabetical index

Solar eclipses are very important to scientists, for it is during eclipses that they are able to observe the **Sun's corona**, which normally is too bright.

Solar eclipses cover a smaller area on the surface of the Earth, so they can be observed only from certain places, in contrast to what occurs with lunar eclipses.

You must **never** observe a solar eclipse **directly**, for that can result in serious damage to the eyes, including **blindness**. You have to use special glasses that protect against radiation, and not just sunglasses or a pane of glass darkened with smoke.

### SOME SOLAR ECLIPSES IN THE DECADE FOLLOWING 2000

| Date | Type |
|---|---|
| May 31, 2003 | annular |
| October 3, 2005 | annular |

# THE SOLAR SYSTEM

Among the thousands of **stars** that make up our **galaxy** there is one medium-sized star located on one edge of the galaxy that has special interest for us because we live near it; it is the **Sun**. This unique star and the planets that revolve around it make up what we call the **solar system**. It is comprised of the nine known planets, their satellites, and a band of rocky remains that form what is known as the Asteroid Belt; it is located between Mars and Jupiter, and it may be what is left over from an old planet that was destroyed.

## THE SUN AND ITS PLANETS

The Sun is larger than any of its **planets**; the largest of them, Jupiter, is just a tenth of its size. In addition, almost 99 percent of the **mass** that constitutes the solar system is concentrated in the Sun, and the rest is spread among the planets. These revolve around the Sun in more or less circular **orbits**. The time they take to complete their respective orbit is known as a **year**; in the case of the Earth, that is about 365 days. But the length of the year is variable, and it depends on how far each planet is from the Sun. The shortest year is **Mercury's**, which lasts just 88 days; the longest is **Pluto's**, which is equivalent to 248 Earth years.

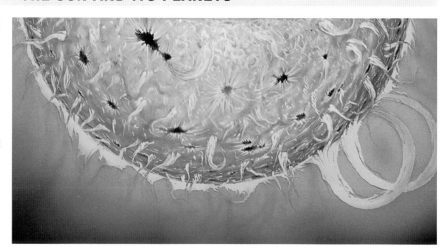

### TYPES OF PLANETS

There are two main groups of planets: the smaller planets such as Mercury, Venus, Earth, and Mars have a rocky structure of fairly high density. The giant planets (Jupiter, Saturn, Uranus, and Neptune) are very large, of low density, and are made up in large measure by matter in liquid or gaseous form. Pluto is the smallest planet, and along with any new planets that may be discovered, it is part of the outer planets.

The force that keeps the planets in orbit around the Sun is the same one that holds satellites in orbit around their planets.

The tremendous heat given off by the Sun decreases with distance. On Mercury the surface temperature reaches 660° F (350° C) during the day, but on Pluto it doesn't get above 433° F (223° C) below zero.

Mercury
Venus
Earth
Mars
Jupiter
Saturn
Uranus
Neptune
Pluto

# ORIGINS

Around five billion years ago what is now occupied by the solar system was filled with dust and gas that evidently came from a **supernova** that had exploded earlier. All this matter began to condense because of the **force of gravity**. In that way a dense core formed that pulled together most of the mass, and that is how the **Sun** was formed. The rest of the matter continued to form a sort of disk around this **star**. The collisions among the particles of dust and the small rocks also caused them to condense around certain points, thereby producing the masses that we refer to as **planets**.

## BODE'S LAW

This astronomer discovered an interesting numerical relationship among the distances from the **planets** to the **Sun**. First there is a numerical series made up of zero, which represents **Mercury**, and three for **Venus**, and the remaining numbers that stand for the planets are double the preceding one. If we add four to each number and divide by ten, we get a new series in which one is the distance from the **Earth** to the Sun and each number coincides almost exactly with the distance of the respective planet from our star.

| | Mercury | Venus | Earth | Mars | Asteroids | Jupiter | Saturn | Uranus | Neptune | Pluto |
|---|---|---|---|---|---|---|---|---|---|---|
| | 0 | 3 | 6 | 12 | 24 | 48 | 96 | 192 | 384 | 768 |
| | 0,4 | 0,7 | 1 | 1,6 | 2,8 | 5,2 | 10 | 19,6 | 38,8 | 77,2 |

## SATELLITES

The solar system is made up of several small systems known as **planetary systems**; they are governed by the same laws that connect the Sun with its **planets**. Every one of these systems is made up of a planet and one or more **satellites** revolving around it. The smaller planets have only a few satellites: **Mercury** and **Venus** have none; **Mars** has two; and the **Earth** has one. The giant planets, on the other hand, have many: **Jupiter**, sixteen; **Saturn**, twenty-three; **Uranus**, fifteen; and **Neptune**, eight. Finally, **Pluto**, like the smaller planets, has just one satellite.

# THE SUN: OUR STAR

The Sun is a modest star located in a corner of our galaxy, and it probably doesn't stand out in the sky that can be seen from some hypothetical planet belonging to another star. Still, we depend on it to live, and because it is so close, we know it better than any other star, and it has been the subject of the greatest number of studies.

## COMPOSITION

The **Sun** is a **star** of the **yellow dwarf** type; in other words, it is relatively modest in size, but because it is located about 93 million miles (150 million km) from Earth, for us it is the most important star in the sky. It is a large sphere made up of about 24 percent **helium**, 75 percent **hydrogen**, and 1 percent other elements. **Nuclear** fusion reactions take place inside the Sun; as a result, hydrogen atoms fuse and produce helium atoms, which are slightly heavier, and they produce a small quantity of energy. This **energy** is what is given off into space and reaches Earth to make life possible.

### THE SUN

| Physical Characteristics | | Astronomical Characteristics | |
|---|---|---|---|
| Surface Temperature | 11,000° F | Visual Magnitude | −26.8 |
| Diameter | 849,443 miles (1,392,530 km) | Absolute Magnitude | +4.8 |
| Volume | $1.41 \times 10^{18}$ m³ | Median Distance from Earth | 9,089,000 miles (149,600,000 km) |
| Mass | $2 \times 10^{30}$ kg | Rotation Period | 25–30 days |

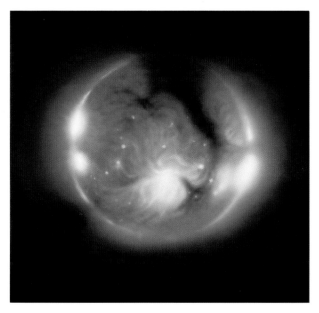

Between 1973 and 1979, *Skylab* provided the first data on solar radiation that doesn't reach the surface of the Earth.

The **diameter** of the Sun is around 100 times greater than that of the Earth.

## OBSERVING THE SUN

**You must never look directly** at the Sun, since that could cause blindness; especially, **binoculars** and **telescopes** must never be used. Special telescopes are used for observing the Sun; they reflect the image on a mirror and are fitted with filters. Receptors located at the bottom of large tanks are used for studying the radiation that comes from the center of the Sun. Important data concerning the structure of the Sun have been gathered by the **satellites**, **spaceships**, and **laboratories** that have been launched into space.

 The Sun weighs around 330,000 times more than the Earth.

 In producing energy, the Sun consumes 600 million tons of hydrogen every second.

Artist's conjecture showing a cross section of the Sun and its different layers

# STRUCTURE

The Sun is an enormous mass of gases made up of a very hot **core** (1) that is surrounded by a succession of cooler layers. In the core the **temperature** is around 36 million degrees Fahrenheit, but the surface reaches just 11,000 degrees. On top of the core there is a **radiant zone** (2), which emits radiation to the outside; this is followed by the **convection zone** (3), where huge columns of gases are formed and rise and fall; and finally there is the surface, known as the **photosphere** (4), which is a thin layer.

On top of the photosphere there is a thin layer called the **chromosphere** (which measures around 1,800 miles/3,000 km in thickness), followed by the **solar corona**, which is a very hot zone. These two zones can be considered the Sun's atmosphere.

## THE CORE

The **core** makes up about a fourth of the Sun's volume.

It takes solar radiation, including visible light, about eight minutes to reach Earth.

The **radiant zone** is a thick layer that extends between 0.25 and 0.80 of the Sun's radius.

The **speed of light** is 182,873 miles (299,792.5 km) per second.

| X-rays | | ultraviolet light | visible light | infrared light | radio waves |

**The spectrum** of solar radiation

The part of solar radiation that we humans can see is called visible light; it includes wavelengths between 380 and 780 mm.

## SOLAR LIGHT

The Sun puts out a great quantity of **electromagnetic radiation** as a result of the nuclear reactions that take place inside it. One part of this radiation is given off toward the outside and reaches our planet. It has very diverse **wavelengths**, ranging from **X-rays** to **radio waves**, and we can see only a part of the range in the form of **visible light**. To our eyes this appears white, but it is also made up of various types of radiation of different wavelengths, each of which is a **color**.

View of the convection zone where the hot gas rises, cools off, and once again descends toward the interior.

# THE SUN: AN ACTIVE STAR

If we contemplate a landscape bathed in sunlight, the light floods everything uniformly. We can't look at the Sun directly without running the risk of going blind. But if we observe it through a **solar telescope**, we discover that its surface is like a great sea, with enormous waves, spots that move, and a brilliant halo that surrounds it.

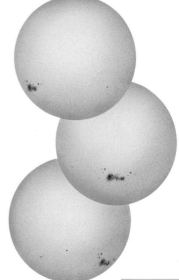

## SUNSPOTS

These are dark-colored areas on the surface of the Sun that are at a lower temperature than the background. They appear in the vicinity of the equator and are never observed at the Sun's poles. They consists of a darker central area and a shaded border that is somewhat lighter in color. Their shape and size are very changeable. They may last a few hours or several months, depending on their size.

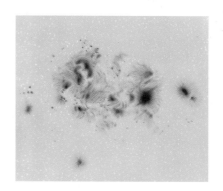

Sunspots move with the rotation of the Sun; they are not constant in number, but vary in an eleven-year cycle that is known as the solar cycle.

The most recent period of maximum **solar activity** was between 1990–1991; the next is expected around the year 2003.

### TEMPERATURE

Sunspots are around 7,200°F (4,000°C).

The greatest solar prominences cause interference on Earth that affects **telecommunications**.

Solar prominences in the form of a loop (**quiescent prominences**) don't achieve much height, but can last for months.

## SOLAR PROMINENCES

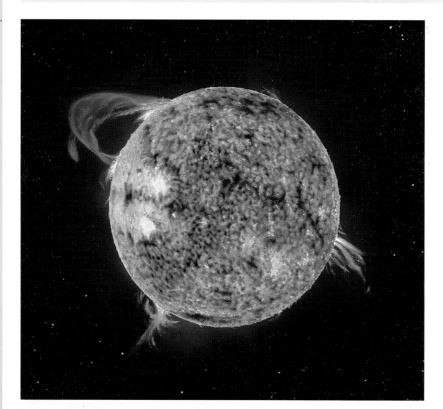

The **chromosphere** is at a temperature of around 180,000 degrees, but it is not very dense, and it hardly gives off any energy. Still, it is the site of a very spectacular phenomenon. This involves huge flare-ups that rise thousands of miles/kilometers above the surface and are referred to as **solar prominences**. These columns of gas extend through the **solar corona** and through space, sometimes traveling 610,000 miles (a million km) from the surface.

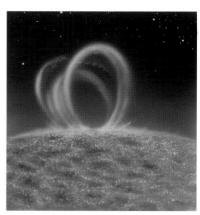

**Eruptive solar prominences** are of short duration, rarely lasting more than thirty minutes.

Solar prominence in the shape of a loop

## THE SUN'S CORONA

This part of the Sun can be considered its **outer atmosphere**; it extends from the **chromosphere** outward several million miles/kilometers into space. This area doesn't have much density at all, and even though it is at a temperature of around 1.8 million degrees Fahrenheit, it gives off very little **radiation**. The shape of the Sun's corona varies, and it also depends on the **activity cycles**; it extends farther into space during years of maximum activity.

The solar corona can be observed during **total eclipses** of the Sun, when the lunar disk entirely covers the solar surface, leaving only the **photosphere** visible, surrounded by a broad, whitish halo with a series of flare-ups and filaments; this is the corona.

The solar corona gives off X-rays and ultraviolet light.

Radiograph of the **solar corona**, the source of particle flow.

## SOLAR WIND

**Solar wind** is the term given to a continuous flow of **particles** that the Sun gives off in all directions into space. This flow is very sparse, containing only four or five particles per cubic centimeter; just the same, when it arrives at Earth, it disturbs **telecommunications** and causes effects as spectacular as the **aurora borealis**. It's also because of solar wind that **comets** form visible tails.

### SPEED

Solar wind moves at around 240 miles/ 400 km per second.

Solar wind is comprised mainly of hydrogen atoms, helium atoms, and free electrons.

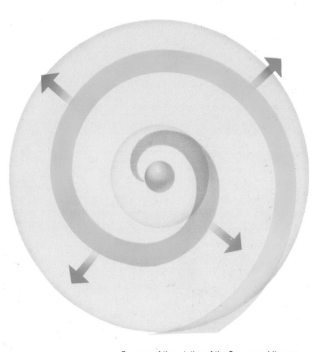

Because of the rotation of the Sun around its own axis, solar wind is given off in a spiral.

# MERCURY

Slightly larger than the **Moon**, this inner **planet** is the smallest one and the closest to the Sun. It has no **satellites**, and its orbit is very elliptical; it is also at a greater angle than that of the other planets with respect to the general plane (**elliptical**). Its surface is full of **craters**; temperatures on Mercury are extremely high, and there is no atmosphere.

## GENERAL CHARACTERISTICS

This small planet is so close to the Sun that its surface temperatures are very high, over 900 degrees Fahrenheit during the day; during the night, though, they are very cold, at least 340 degrees below zero. This happens because, due to its size, it has no **atmosphere:** its **gravity** is too weak to hold onto the type of gassy covering that we have here on Earth.

A spaceship needs a speed of 2.6 miles per second to blast off from its surface.

Gravity on the surface of Mercury is .39 times that of Earth.

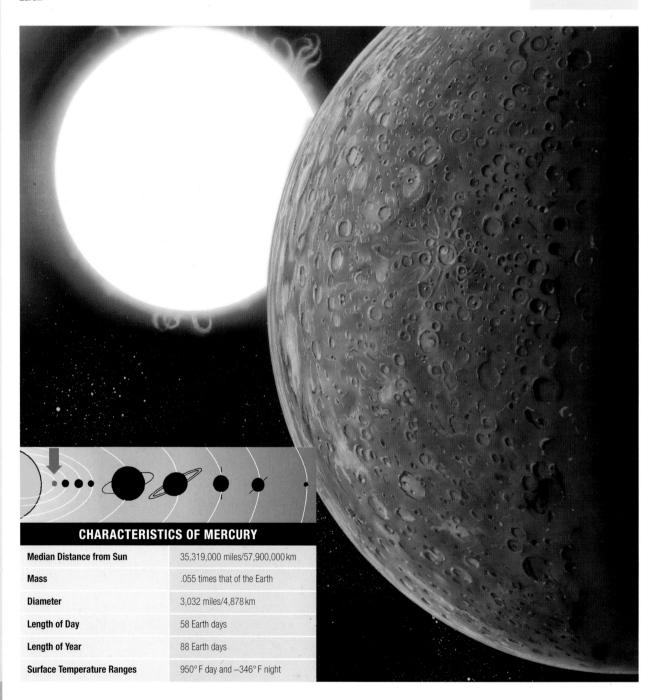

### CHARACTERISTICS OF MERCURY

| | |
|---|---|
| **Median Distance from Sun** | 35,319,000 miles/57,900,000 km |
| **Mass** | .055 times that of the Earth |
| **Diameter** | 3,032 miles/4,878 km |
| **Length of Day** | 58 Earth days |
| **Length of Year** | 88 Earth days |
| **Surface Temperature Ranges** | 950° F day and −346° F night |

## STRUCTURE OF THE PLANET

Mercury is a solid planet made up of a metal **core** (1) covered by a **mantle** (2) of rocks; the **crust** (3) is located on top of this. Mercury was formed some 4.5 billion years ago, and because of its small size, it solidified quickly. It is believed that there was never any **volcanic activity** on the planet, so it has not experienced any changes since its birth, except for impacts of **meteorites**.

Mercury has a very weak magnetic field.

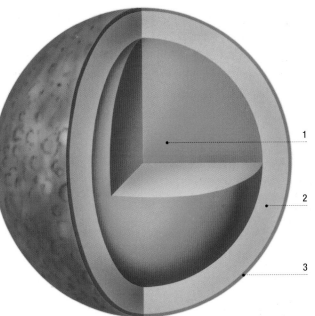

Cross-sectional view of the planet

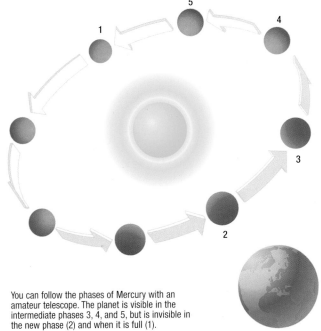

You can follow the phases of Mercury with an amateur telescope. The planet is visible in the intermediate phases 3, 4, and 5, but is invisible in the new phase (2) and when it is full (1).

## OBSERVING MERCURY

It is difficult to observe this small planet from Earth because of its orbit and its proximity to the Sun. It is visible close to the horizon during the dawn and dusk, when it is bathed in light. The best time to observe it with the aid of a telescope is when it is in the stages of **waxing** or **waning**; when it is **full**, it passes behind and is obscured by the Sun, and in its **new phase** it is scarcely visible.

It appears that there are vestiges of **ice** in the polar regions of the planet, but no one credits the possibility that it harbors any **life**.

## A LUNAR SURFACE

No one knew just what the surface of this planet was like until 1974, since it can't be observed in detail using telescopes. It was presumed, though, that the surface was heavily cratered. In that year the spaceship *Mariner X* passed near the planet and took many photographs that revealed that Mercury's landscape is similar to that of the Moon, with numerous craters caused by impacts from **meteorites**, plus some crevasses, which also are probably due to those collisions. The soil is grayish in color.

## LARGEST CRATER

The largest of Mercury's craters is the one designated *Fossa caloris*, which measures some 800 miles (1,300 km) in diameter and is surrounded by undulating terrain.

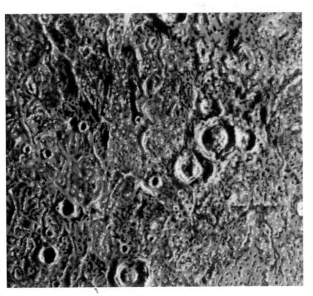

Detail of the planet's surface

# VENUS

This is the planet that is closest to Earth in size, and is one of the brightest ones visible in the sky at dawn and dusk. Until the beginning of the Space Age, practically nothing was known about its topography because of the **dense atmosphere** that surrounds it. Its surface reaches very high temperatures. This planet has no **satellites**.

## GENERAL CHARACTERISTICS

Venus is an **inner planet** with dimensions and a mass similar to those of the Earth; before it became possible to study its surface, it was thought to be covered with seas. However, spaceships have revealed that the soil is completely dry, and that temperatures there reach over 800 degrees Fahrenheit; that is an obstacle to any kind of life. One notable characteristic of this planet is that it rotates in a direction opposite that of the other planets in the solar system. It seems that the reason for that is the impact of a large **meteorite**. In addition, this is the only planet whose **day** is longer than its **year**.

Gravity on the surface of Venus is .88 that of Earth.

The speed required for a spaceship to leave the surface of Venus is 6.3 miles (10.4 km)/sec.

### CHARACTERISTICS OF VENUS

| | |
|---|---|
| **Median Distance from Sun** | 65,880,000 miles/108,000,000 km |
| **Mass** | 0.81 times that of the Earth |
| **Diameter** | 7,519 miles/12,102 km |
| **Length of Day** | 243 Earth days |
| **Length of Year** | 225 Earth days |
| **Surface Temperature** | 866° F/480° C |

## STRUCTURE OF THE PLANET

Venus is a solid planet made up of a **core** (1) of a metallic nature, surrounded by a very thick rocky **mantle** (2), and a thin **crust** (3). **Space probes** that have succeeded in landing on the surface have shown a landscape with broad plains, deep valleys, and mountains exceeding 32,800 feet (10,000 m) in height. The deep depressions seem to be the remains of **seas** that must have existed before the present thick **atmosphere** was formed.

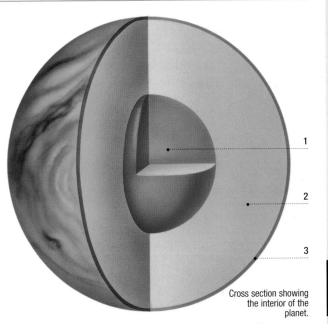

  There are active volcanoes on Venus.

The clouds on Venus move at high speed, producing huge storms at that altitude.

 The pressure recorded on the surface reaches 1,470 p.s.i./100 atmospheres.

1

2

3

Cross section showing the interior of the planet.

## EXPLORING VENUS

The first spaceship to land on the surface was the Russian *Venera 7* in 1970. *Mariner X* photographed the surface of Mercury and the atmosphere of Venus in 1973. In 1978 the spaceship *Venus Pioneer* used parachutes to launch a probe when it passed by the planet. In 1992 the probe *Magellan* mapped the surface. All these spacecraft have allowed us to get to know the terrain that is hidden underneath the planet's atmosphere, which had always been an impenetrable barrier.

## THE ATMOSPHERE OF VENUS

The atmospheric layer of this planet is made up principally of **carbon dioxide**, along with some other elements and chemical compounds. Above the surface, at an altitude of thirty to forty-six miles (fifty to seventy-five km), there is a dense layer of **clouds** that completely surrounds the planet. It is the location of **intense electrical storms**. This cloudy layer acts as a screen and causes a major **greenhouse effect**; in other words, it keeps heat from escaping. That explains why the surface reaches 866° F/480° C, making the existence of life an impossibility.

 The **atmosphere** of Venus consists of 96 percent **carbon dioxide**; the rest is made up of **nitrogen**, **water vapor**, **sulfur dioxide**, and lesser quantities of other chemical compounds.

### CALM WINDS

Winds on the surface of the planet are mild, and since there are no clouds, the tops of the mountains are easily visible.

The Russian probe *Venera* on the surface of Venus.

# THE EARTH: A SPECIAL PLANET

Of all the planets that make up the **solar system**, the Earth is the only one that supports **life**. This is a planet where the surface temperatures remain moderate because of the presence of **water** and an **atmosphere**. The **crust** remains very active geologically and is in a continual state of formation. The Earth has just one **satellite** (the Moon).

## GENERAL CHARACTERISTICS

Since this planet is the one where we live, we use it as a point of reference in studying the remaining heavenly bodies. That is why many characteristics of the other planets are measured with respect to the Earth, as in the case of **mass**, **gravity**, and the duration of **rotational periods**. Even though this planet is our home, there are many features about it that are not well known. Starting at the end of the twentieth century, **spaceships and space stations** have made it possible to study the Earth from afar and clear up some existing mysteries.

**Gravity** on the surface of the Earth is 9.81 m/s$^2$.

The speed a spaceship needs to leave the surface of the Earth is 6.8 miles (11.2 km)/sec.

The Earth as seen from the Moon.

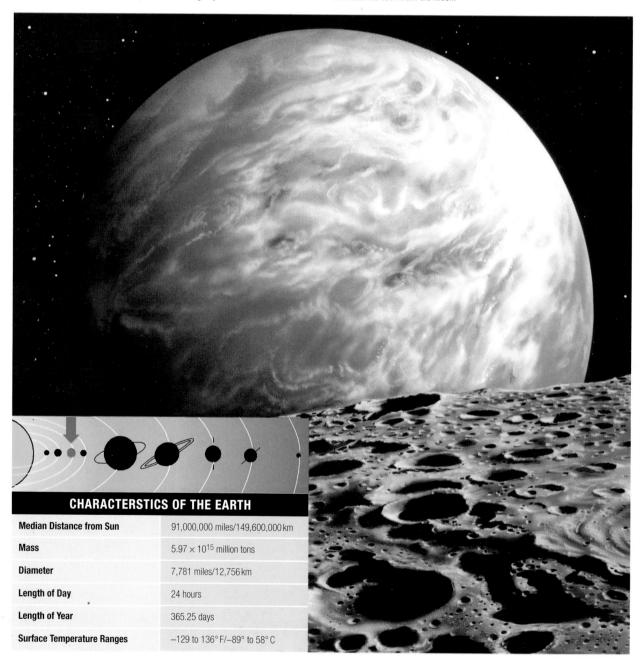

| CHARACTERSTICS OF THE EARTH | |
|---|---|
| Median Distance from Sun | 91,000,000 miles/149,600,000 km |
| Mass | 5.97 × 10$^{15}$ million tons |
| Diameter | 7,781 miles/12,756 km |
| Length of Day | 24 hours |
| Length of Year | 365.25 days |
| Surface Temperature Ranges | −129 to 136° F/−89° to 58° C |

## ORIGIN OF THE EARTH

Our planet was formed at the same time as the Sun, when the great mass of matter that resulted from the explosion of an earlier star (a **supernova**) condensed. Most of this matter went into forming the **Sun**, but the rest of it went into the formation of the planets when some clusters of larger rocks attracted the dust and smaller materials that were around them. This process took some 100 million years. The newly formed **planet** heated up a lot when the radioactive materials it contained disintegrated and melted. Then there began a cooling-off process in the outer layers; they ended up solidifying and produced the **crust**.

The newly formed Earth had no atmosphere and suffered numerous collisions from rocky fragments that caused it to heat up.

Four phases in the formation of the Earth: Bombardment by meteorites (1) heated it up to the melting point (2); at that time the heaviest elements sank into the Earth (3) to form a core inside the planet (4).

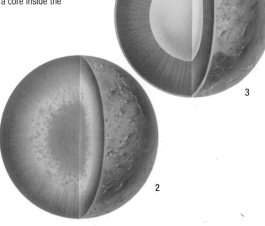

When the Sun turns into a red giant, its greater proximity will draw off the gases that make up the Earth's atmosphere and leave the Earth without any atmosphere at all.

## THE END OF THE EARTH

The destiny of our planet depends on the evolution of the **Sun**. When this star has consumed all its **hydrogen** and **helium** reserves, it will turn into a red giant; it will then increase considerably in size. The nearest planets such as Mercury and Venus will be engulfed in the tremendous solar mass, and the heat will be so intense on our planet that every form of life will disappear. The surface of the Earth will be transformed into a desert, and the ocean basins will be totally empty; the landscape will be similar to what we can observe on Mars.

### DEPENDING ON THE SUN

Since the Sun has adequate reserves for another five billion years, our planet still has that much time to live.

# THE EARTH: COMPOSITION

The Earth is one of the **solid planets**; it is made up of several layers of varying nature that surround an incandescent core. The **crust** is the outermost of these layers, and is also the one where life exists.

The crust is not a static structure, but one that is in constant movement and subject to many phenomena (volcanoes, earthquakes, and so forth).

## A CROSS SECTION OF THE PLANET

Our planet is one of the solid planets (Mercury, Venus, Earth, and Mars); it is small in comparison to the giant planets (Jupiter, Saturn, Uranus, and Neptune) that are made up of gases. The innermost part of the Earth is the **core**, which is made up primarily of **iron** and **nickel**; it consists of a central solid part (1) and a liquid outer one (2). That is followed by the **mantle** (3), which is solid in its deepest part but then becomes viscous;

lastly there is the **crust** (4), which is called the **lithosphere**; it is the thinnest layer, and is comprised of solid material.

On top of the crust there are the oceans and seas (which make up the **hydrosphere**) and the **atmosphere**, which is the air that we breathe.

The core has a radius of 2,135 miles (3,500 km) and contains lots of **siderite** ($FeCO_3$).

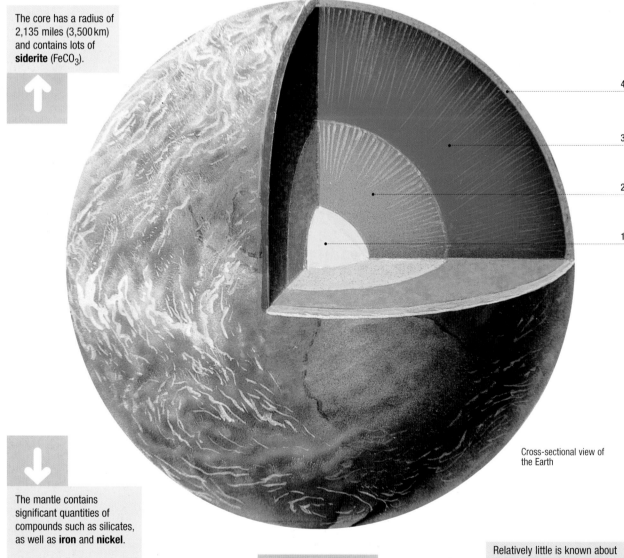

Cross-sectional view of the Earth

4
3
2
1

The mantle contains significant quantities of compounds such as silicates, as well as **iron** and **nickel**.

The mantle is about 1,524 miles (2,500 km) thick, and the fluid part occupies the uppermost 122 miles (200 km) near the crust.

### SIMA AND SIAL

The crust is about 42 miles (70 km) thick. The deepest part is known as **sima**, and the **sial** is situated on top of that (it is given that name because it contains a large amount of aluminum).

Relatively little is known about the interior of the Earth. The greatest distance ever ventured into the Earth is only about 9 miles/15 km.

# THE EARTH'S CRUST

The outermost layer of our planet is not uniform, but quite varied in thickness. It is thinner in the bottom of the oceans and thicker in the continents, but this difference is compensated for with density: In the thinnest areas, the materials are denser than in the areas of greater thickness. That way there is a balance, and the entire crust weighs about the same in all places.

→ The distribution of light materials in the thick layers and the dense ones in the thin layers is referred to as **isostasy**.

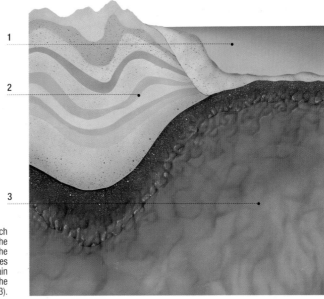

The Earth's crust is much thinner in the bottoms of the oceans (1) than on the continents (2), and the plates of the crust that contain the continents float on the mantle (3).

## ACTIVITY IN THE CRUST

**Volcanoes** are the places where the incandescent material of the Earth's mantle finds a way to the surface through the crust. **Lava** comes out of the mouth of the **crater** and eventually cools off on the surface to form a very light rock known as **pumice stone**.

In addition, the continents, which are floating on top of the mantle, are slowly changing position. So America and Europe have been growing further apart for millions of years, and as a result the Atlantic Ocean keeps getting wider.

When melted rock or magma (1) rises to the surface (2) through a crater, it produces a volcano (3) as successive layers (4) of cooled lava accumulate.

↓ The areas that have lots of volcanoes are also the ones where there are lots of earthquakes.

### CONTINENTAL DRIFT

The movement of the continents is called **continental drift**. Some parts of the continents are rising while others are sinking because of the push from the continental masses.

# EARTH: THE BLUE PLANET

Astronauts have always described the Earth as the blue planet. The oceans and the gases in the atmosphere are responsible for this color; they are the outer components of the crust that make life possible. The coverings of water and air are unique in the whole solar system.

## THE HYDROSPHERE

All the **water** that exists on the surface of our planet is known as the **hydrosphere**; this includes the **oceans** and **seas**, the rivers, lakes, wetlands, glaciers, the poles, and so forth. These features formed from the **vapor** produced in **volcanic eruptions** in the course of millions of years, when eruptions were much more frequent than they are today. That vapor condensed in the form of clouds that subsequently produced rain. The majority of the Earth's water is contained in the oceans, which cover nearly three-quarters of its surface.

### LOTS OF SALT WATER

The Earth contains 845,450,600 cubic miles (1,385,984,610 cu km) of water. The salt water of the seas comprises 96.54 percent of all the water on the planet.

Life as we know it can survive only on planets where there is water in liquid form.

Some other planets contain water, but only in the form of ice.

As seen from space, the Earth stands out because of the blue color imparted by the seas and the atmosphere.

## THE SEAS AND OCEANS

The depressions contained between the continents form the **oceans**; each ocean has smaller areas that are the **seas**. There are three large oceans: the **Atlantic**, the **Pacific**, and the **Indian**. The bottoms of the oceans are not smooth, but are crossed by large mountain ranges. Some summits stick out of the water and form **ocean islands**, such as the Canaries and the Azores. There are great **currents** in all the oceans; some are cold, and others are warm. They are very important to the climate of the regions to which they flow. Water helps in minimizing the temperature differences at various hours of the day and at different times of the year.

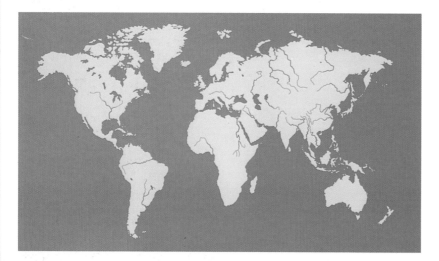

In the **Northern Hemisphere** the seas occupy about 94 million square miles (154 million sq km), and the land some 61 million square miles (100 million sq km).

In the **Southern Hemisphere** the seas occupy some 126 million square miles (206 million sq km), and the land some 29 million square miles (48 million sq km).

## THE ATMOSPHERE

At first the planet had a different **atmosphere** than it has at present. When volcanic eruptions emitted **water vapor** that was deposited in the form of rain, the lowest areas of the planet began to fill up with water and form the oceans. The first **plants** appeared in the oceans; they began to make **oxygen**, and when they released it they transformed the initial, unbreathable atmosphere into the oxygen-rich one that we have today.

## AN ENVIRONMENT IN CONSTANT CHANGE

Meteorological maps show us the constant changes to which the **atmosphere** is subject. Cloud fronts are pushed along by winds, producing rain and sunshine and raising and lowering temperatures. But this happens only in the lower layers, the ones that affect us most directly. As you gain altitude, there is less **oxygen** and the temperature drops.

## ATMOSPHERIC LAYERS

From the ground, the atmosphere has several layers: up to 6 miles (10 km) in altitude there is the **troposphere**, which is where we can live; up to 60 miles (100 km) in altitude there are the **stratosphere** and the **mesosphere**, and from there up to 600 miles (1,000 km) the area that produces the aurora borealis (the **thermosphere** and the **exosphere**). The outer area continues out to 6,000 miles (10,000 km), and it contains a belt of radiation (the **magnetosphere**).

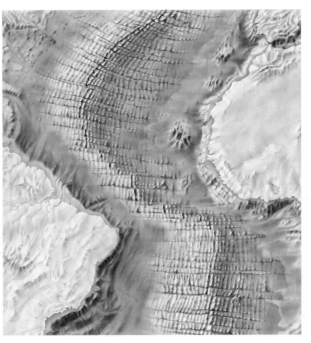

What the Atlantic Ocean would look like if it had no water.

 The atmosphere acts like a protective shield against lethal radiation and meteorites.

The protective **ozone** layer is located 24 miles (40 km) over the surface of the Earth.

# THE EARTH IN MOTION

The Earth is in constant motion. Along with the rest of the planets, plus the Sun, it moves throughout our galaxy, but that doesn't affect us in our daily lives. It is of greater significance for us that it spins on its axis, producing day and night, and that it revolves around the Sun, making the seasons in the temperate latitudes.

## ROTATION ON ITS AXIS

If we look at the night sky for several hours, we see that the stars move around a fixed point; in the Northern Hemisphere, this point is the **North Star**. Later the day dawns and the Sun shines. All this is the result of the Earth's rotation on its axis.

The **constellations** of stars vary throughout the year because in addition to rotating on its axis, the Earth is moving around the **Sun**. Every two months the same constellations are in the sky.

By taking a photograph of the night sky with a very long exposure time, we get this type of image. Every line is produced by a star or other celestial body as it moves around the point of revolution.

The rotational speed at the equator is 1,016 miles per hour (1,665 km/h).

## DAY

**Day** is one basic unit of time. It is the time it takes for the Earth to rotate around its axis, so that a spectator located at a specific point looks at a series of heavenly bodies in the sky until they are repeated. Since a complete rotation equals 360 degrees, every hour the planet moves 15 degrees.

### ARE WE MOVING?

We don't perceive the movement as the Earth rotates on its axis and around the Sun because we are all traveling at the same speed. It is similar to traveling by car or plane. We perceive the speed only when we look out the window.

In the areas near the Earth's poles, there are several months when there is only night, and others when it's always day. This is due to the angle of inclination of the Earth's rotational axis.

The rotational speed of a point located on the pole (through which the rotational axis passes) is zero miles/kilometers per hour.

The trajectory of the Sun over the horizon is different every day of the year.

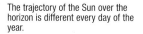

## THE SEASONS

The phenomenon of the **seasons** involves the changes in the length of the day and the angle of the Sun's rays at different times of the year in many parts of the planet, except for the regions of the equator. This produces shorter days and lower temperatures in one period of the year, which we know as **winter**, and other days that are longer and higher in temperature during the time known as **summer**. This effect is due to the fact that the Earth's axis is tipped 23 degrees with respect to the plane in which it moves around the Sun.

In the **winter** the Sun passes at a lower height in the sky and for a shorter time, so it provides less heat.

In the **summer**, the Sun passes at a greater height in the sky and is there for a longer time, so it provides more heat.

## SOLSTICES AND EQUINOXES

**Solstices** occur on December 22 or 23 (winter solstice in the N. Hemisphere, summer solstice in the S. Hemisphere), and on June 22 or 23 (summer solstice in the N. Hemisphere, winter solstice in the S. Hemisphere).

When the inclination of the Earth's axis causes the Northern Hemisphere to point toward the Sun, it is **summer** in that Hemisphere and winter in the Southern Hemisphere. The moment when the inclination is most pronounced is called the **solstice**—the summer solstice in the Northern Hemisphere, and the winter solstice in the Southern. At the opposite point in the orbit the situation is repeated, but in reverse.

The intermediate positions are known as the autumn or spring **equinoxes**, depending on the hemisphere. At the equinoxes, day and night are of equal length; but starting at that date, they begin to increase (in the case of the **spring equinox**) or decrease (in the case of the **autumn equinox**).

**Equinoxes** occur on March 21 or 22 (spring equinox in the N. Hemisphere, autumn equinox in the S. Hemisphere), and on October 21 or 22 (autumn equinox in the N. Hemisphere, spring equinox in the S. Hemisphere).

In its orbit around the Sun, the axis of the Earth always points in the same direction, toward the North Star.

# OUTER SPACE AND THE EARTH

The Earth is not a planet isolated in space, but rather is subject to the influences of other heavenly bodies, mainly the Sun. That star gives off a great amount of radiation that reaches our planet in very diverse forms, such as radio waves, visible light, heat, and others. The Moon also exerts an influence on the Earth's tides.

## IMPACTS OF METEORITES

**Meteorites** are fragments of rock that come from space and reach our planet at high speed. Most of them are small, and as they pass through the atmosphere they are burned up completely and converted to gases. Some of them are large enough for a small solid portion to reach the Earth's surface. They then become buried in the soil and are discovered only by accident. Still, there are others that are large enough to produce a hole similar to a **crater** on the Moon.

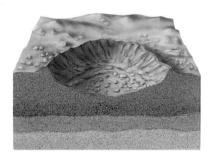
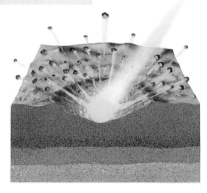

A meteorite is a rock that comes from space. Here it is shown crashing on the Earth's surface and leaving a crater.

## THE EFFECTS OF METEORITES

Crater in the Arizona desert, caused by a large meteorite.

Most meteorite impacts go unnoticed. Only a few of them have an appreciable effect on the Earth. One of the largest craters still in existence is in the **Arizona** desert. One of the theories on the disappearance of the **dinosaurs** holds that a giant meteorite collided with the Earth and produced the disaster. It caused an enormous cloud that lasted for several years, blocking the sky and lowering temperatures so much that most plants and giant reptiles abruptly disappeared.

## TIDES

The **force of gravity** from the **Moon** is the major cause of the periodic rise and ebb in the levels of the sea. The Moon exercises an attraction on the mass of water facing it, producing a **high tide** in that part of the planet, and a **low tide** on the opposite side. Also, when the Moon and the Sun are lined up, their attractions combine to produce exceptionally high tides that are known as **spring tides**.

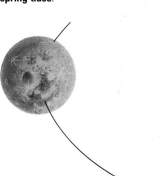
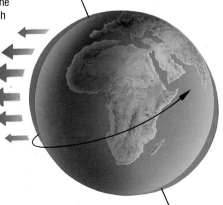

# SOLAR WIND

The flow of protons and electrons from the Sun causes major alterations in radio and television communications during the times of greatest intensity. Every ten or eleven years, solar activity reaches a peak, and at those times there are electromagnetic storms that can momentarily interrupt all telecommunications. These **electromagnetic storms** also affect meteorology.

Magnetic North

Geographic North Pole

Van Allen's Belts

 Van Allen's Belts are areas that retain particles of solar wind due to the Earth's magnetic field.

The magnetosphere is the part of the Earth's magnetic field that extends outward toward space.

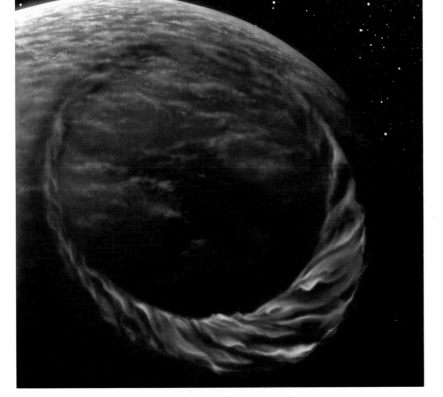

## POLAR AURORAS

This phenomenon is one of the most impressive ones caused by the arrival of charged particles in the Earth's atmosphere; however, the aurora can be observed only near the polar regions. Since people in the Northern Hemisphere live in latitudes near the pole, this phenomenon is known as the **aurora borealis**, even though it also occurs in Antarctica. The auroras take place when solar wind reaches the Earth's atmosphere and the electrons and protons emit beautifully colored lights.

From space, the polar aurora looks like a circle of light around the pole.

## A WONDERFUL SPECTACLE

For an observer on the ground, the polar auroras appear to be huge curtains of beautiful, colored lights that move in the sky.

# THE MOON: OUR SATELLITE AND ITS CONQUEST

Our satellite is one of the main celestial bodies that pass through the Earth's skies; its gravity also exerts an influence on our oceans. The Moon is a **solid** satellite; it is relatively large, and it rotates around us in a cyclical fashion in such a way that part of its surface is always hidden from the Earth. The Moon is also the first heavenly body besides the Earth that people have visited.

## GENERAL CHARACTERISTICS

Our satellite is very different from the Earth. Because of its smaller size, its **gravity** is not strong enough to hold onto an **atmosphere**, so it has no protective layer, and its surface is constantly being bombarded by **meteorites** of all sizes. This is what gives it the characteristic appearance of being covered with **craters** of all sizes and large **plains**.

| CHARACTERISTICS OF THE MOON | |
|---|---|
| **Maximum Distance from Earth** | 247,355 miles / 405,500 km |
| **Minimum Distance from Earth** | 221,613 miles / 363,300 km |
| **Mass** | .0123 times that of Earth |
| **Diameter** | 2,120 miles / 3,476 km |
| **Density** | .62 times that of Earth |
| **Surface Temperature Ranges** | 266° F day and −292° F night |

**Gravity** on the Moon is equivalent to .166 that of the Earth.

The speed required for a spaceship to escape from the surface of the Moon is 1.45 miles (2.38 km)/sec.

Here are a couple of the rocks brought back by the astronauts who explored the lunar surface.

## STRUCTURE

Since the **density** of the Moon is 3.42 with respect to water (the Earth's is 5.52), the inner part of heavy materials must be relatively small. These materials, mainly **iron**, make up the **core** (1). It is surrounded by a **mantle** (2) of melted rock similar to that of the Earth; over that there is the **crust** (3). This is covered by a layer of residue in the form of dust and rocks.

### WHAT IS REGOLITH?

The layer of dust and rocks on the lunar surface is known as **regolith**.

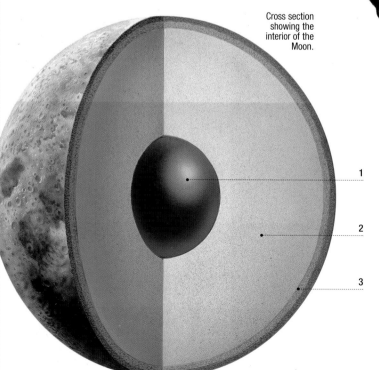

Cross section showing the interior of the Moon.

1

2

3

Because of the Moon's gravity, a person who weighs 154 pounds on Earth weighs only 25.5 pounds on the Moon.

The layer of dust that covers the lunar surface is due to collisions of meteorites from space.

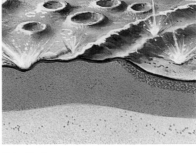

The Moon when it was first formed.

Meteorite shower.

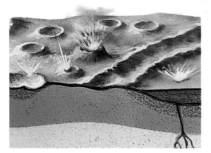

Intense volcanic activity three billion years ago.

Present appearance.

## EVOLUTION

There are three theories concerning the origin of the Moon: 1) When the Moon passed close to the Earth, it was captured and held in orbit; 2) the two celestial bodies originated from the same mass of raw material that was revolving around the Sun; 3) the Moon originated as a bulge in the Earth and was dislodged by centrifugal force.

Currently there is a new theory that is a mixture of these three: When the Earth was forming, it collided with a large space object, and as a result, when part of its mass was expelled, it clumped together and formed our satellite. This theory explains the compositional differences between the Earth and the Moon.

A waning Moon is illuminated on the left side.

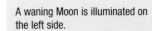

A waxing Moon is illuminated on the right side.

Appearance of the Moon in its four phases.

## THE PHASES OF THE MOON

The Moon moves around the Earth in the same direction that our planet follows. It takes twenty-nine days to complete a single **rotation**, and its illumination by the Sun varies throughout its orbit. When the visible face is illuminated by the Sun, we speak of the **phase of the full moon**; when the Sun illuminates the far side of the Moon, we say that it is in the new moon phase. The **waning** and **waxing** phases are when only half the surface is visible.

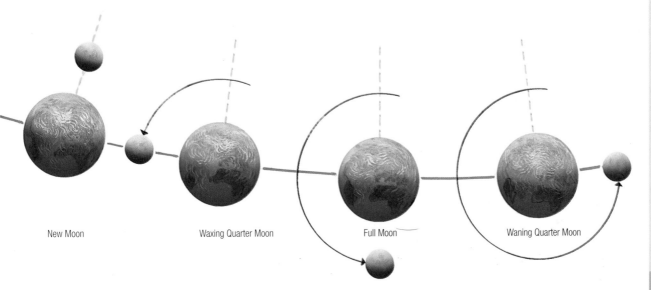

New Moon       Waxing Quarter Moon       Full Moon       Waning Quarter Moon

# THE GEOGRAPHY OF THE MOON

The surface of our satellite as seen from the Earth with simple binoculars shows a landscape of clear, dark spots and outlines of craters. Since the lunar day is the same length as a day on Earth, we always see the same side of the Moon. Man-made satellites have made it possible to explore the far side of the Moon, which is also made up of seas and craters.

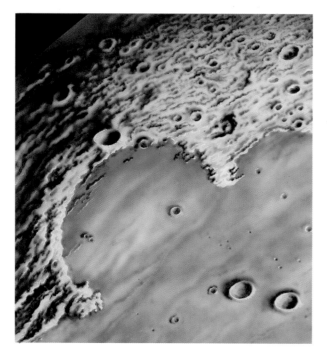

## LUNAR SEAS

These are vast expanses that appear darker in color than the rest of the surface; it used to be thought that these were areas where seas once existed. In fact, they are cavities produced by ancient impacts of large **meteorites** that became covered by a thick layer of lava during the Moon's phase of intense volcanic activity.

### TRANQUILITY

The first astronauts on the Moon walked in the **Sea of Tranquility**.

Some of the most famous seas of the Moon are: The Seas of
**Tranquility**
**Serenity**
**Rains**
**Crisis** and
**Cold**

## CRATERS

There are many **craters** that cover the entire surface of the Moon; they are due to impacts of **meteorites** that collide with full force, since there is no protective atmosphere to burn them up or slow their fall. Their size varies from a few inches/centimeters or less to over 122 miles (200 km) in diameter. Many of these craters have a small elevation or a peak in the middle. Lunar **volcanoes** have also caused some craters, but they never have a peak in the middle.

One of the largest lunar cirques is **Clavius**, which has a diameter of 142 miles (233 km).

### ALWAYS THE SAME SIDE

Since the Moon takes the same amount of time to complete a revolution on its axis as a complete rotation around the Earth, the same side always faces our planet.

Introduction

Space

The solar
system

The Sun

Mercury

Venus

**Earth**

Mars

Asteroids

Jupiter

Saturn

Uranus

Neptune
and Pluto

Exploring
the universe

Astronautics

Alphabetical
index

## THE VISIBLE SIDE OF THE MOON

The seas occupy around 40 percent of the visible surface. The rest is made up of craters, mountain chains, and valleys, as well as a few other structures in the shape of stars; these are due to the impact of material spread by a collision from a meteorite. The lunar relief is very rough, and some mountains are 26,400 feet / 8,000 m high.

The best time to observe the lunar relief with binoculars or a small telescope is in the waxing and waning phases.

Because of the small irregularities in orbits, it is possible to observe some 60 percent of the lunar surface from Earth— in other words, a little more than one entire side of the Moon.

## THE FAR SIDE OF THE MOON

The structure of this side is similar to the visible side, but its geography is very different: It is made up almost entirely of **craters**, and the only existing **seas** are located in the region near what we call the visible side of the Moon. In addition, the thickness of the crust on this side is greater.

### THE UNSEEN SIDE

The first images of the far side of the Moon were obtained in 1959 thanks to the photographs sent back to Earth by the Soviet space probe *Luna 3*.

 The Moon experiences earthquakes that are detectable from Earth.

# MARS

This planet is one of the most noticeable ones because of its reddish color. It is smaller than the Earth; its surface is covered with large rocky and sandy plains, with mountains and small craters. It has a very thin **atmosphere**, and there are considerable expanses of **ice** in the polar regions that vary with the time of year. The planet has two small **satellites**.

## GENERAL CHARACTERISTICS

The planet Mars, whose radius is half that of the Earth, is characterized by the intense red color of its surface, which is formed by **metallic oxides**, and by the white spots of **ice** that cover the polar regions. Since its rotational axis is inclined much the same as the Earth's, the planet has several distinct **seasons** throughout the year.

The surface of Mars, visited by the *Viking* probe.

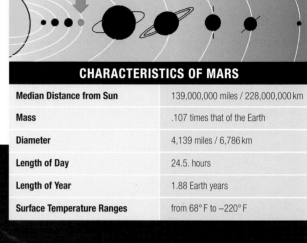

| CHARACTERISTICS OF MARS | |
|---|---|
| **Median Distance from Sun** | 139,000,000 miles / 228,000,000 km |
| **Mass** | .107 times that of the Earth |
| **Diameter** | 4,139 miles / 6,786 km |
| **Length of Day** | 24.5. hours |
| **Length of Year** | 1.88 Earth years |
| **Surface Temperature Ranges** | from 68° F to −220° F |

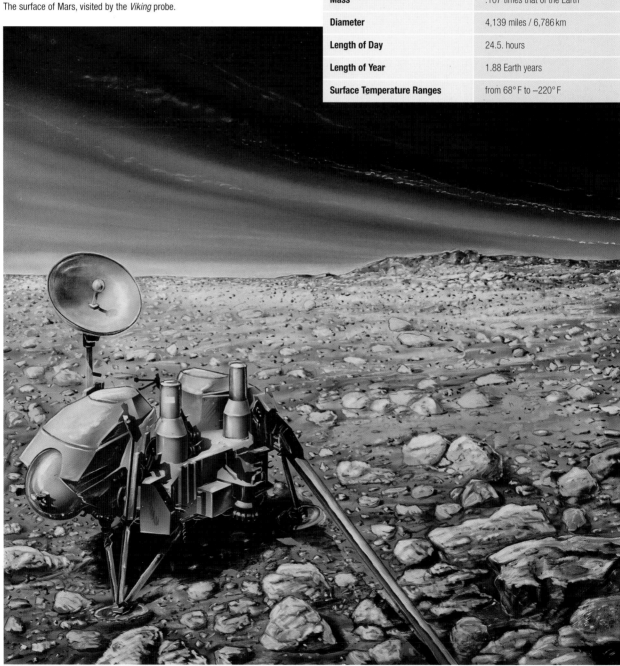

## STRUCTURE OF THE PLANET

Mars is a solid planet made up of a metallic **core** (1) surrounded by a rocky **mantle** (2) and an **outer crust** (3). The relief is characterized by variety. There are vast desert expanses covered with reddish sand and rocks, as well as high **mountains** several times higher than Mount Everest, tectonic valleys of huge size, volcanic **craters**, and some smaller craters caused by the impacts of meteorites.

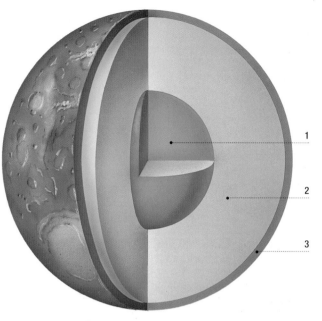

1

2

3

Cross section showing the inside of the planet.

**Mount Olympus** is an ancient volcano; it is the best-known peak on Mars. It is 82,500 feet / 25,000 m high.

Gravity on the surface of Mars is .38 times that of Earth; the speed a rocket needs to reach to escape from its surface is 3 miles (5 km)/sec.

## STORMS

One of the characteristics of this planet is the great **storms** that take place and stir up tremendous quantities of dust. Still, since the atmosphere is very thin, it doesn't take this dust long to settle back onto the surface. This seems to be the explanation for the "**canals**" that the ancient astronomers observed; many thought they were created by inhabitants of the planet.

Strong dust storms are responsible for surface erosion on Mars.

## THE SATELLITES OF MARS

Mars has two small moons that were discovered in 1877; they were christened *Phobos* and *Deimos* after the names of the horses that pulled the chariot of Mars, the Roman god of war. These satellites move slowly through the Martian sky and shine like stars. They are irregular in shape, and it is believed that they are two **asteroids** that were captured by the planet's **force of gravity**. They can be seen only with the aid of a strong telescope.

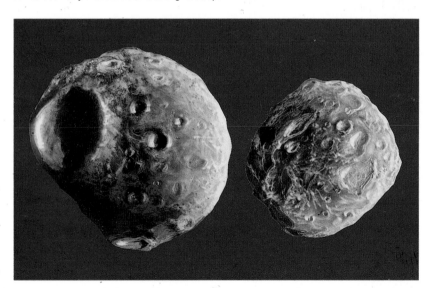

The satellites of Mars

## A THIN ATMOSPHERE

The ice of the polar caps is made up of water and carbon dioxide; the latter is also the main component of the planet's thin **atmosphere**.

**Phobos** measures 16.5 miles (27 km) in diameter; it has a crater 6 miles (10 km) in diameter, and is located 5,700 miles (9,400 km) from the surface of Mars. Its orbit takes seven hours and thirty-nine minutes.

**Deimos** is 10 miles (16 km) in diameter; it is located 14,000 miles (23,000 km) from the surface of Mars. It takes thirty hours to complete its orbit.

# MARS: AN INHABITABLE PLANET?

## MARTIANS

In 1877 **Schiaparelli,** an Italian astronomer, discovered some long, dark lines on the surface of the planet and called them **canals**. That began the speculation on the origin of these features, and many people suggested that they might really be artificial canals constructed by **Martians**. The hypothesis became very popular, and many people thought that life must exist on the planet, or that it once did. This was the start of the belief in Martians, or inhabitants of Mars.

Martians were a favorite theme in science fiction novels and movies until the **space probes** showed that there is currently no life on the planet.

***The War of the Worlds*** (1898), by the British writer **H. G. Wells**, is one of the most famous works of science fiction; it tells of the invasion of the Earth by Martians.

Sometimes the movies have come up with some fantastic appearances for the nonexistent Martians. This photo shows a scene from the film *Mars Attacks!*

## A COLONY ON MARS?

Of all the planets in the solar system, the one that has the best conditions for a possible human colony is Mars. Still, this would require much effort, since, among other things, it has no breathable **atmosphere**. Nevertheless, the exploratory spaceships that have analyzed the surface indicate that there is a large amount of oxygen; it is not available in the air, but chemically bonded to the compounds in the soil, and it could be utilized. Some ancient river beds have also been found, so in the past there may have been life.

The creation of a **colony** from Earth on Mars would require the development of very powerful rockets, and it probably would be done in stages using space stations as intermediate posts for transporting all the necessary materials to set up a permanent base.

It is believed that most of the **water** that existed on Mars is frozen in the subsoil.

Even though the atmosphere on Mars is very thin, it keeps the surface temperature from exceeding 68° F (20° C).

### A METEORITE

Organic chemical compounds and structures similar to those in **microfossils** have been found in a **meteorite** from Mars.

# THE ASTEROID BELT

Between Mars and Jupiter there is a band about 336 million miles (550 million km) wide that contains a countless number of small celestial bodies, rocks of different sizes, and smaller fragments; they are all known as asteroids.

Introduction

Space

The solar system

The Sun

Mercury

Venus

Earth

Mars

Asteroids

Jupiter

Saturn

Uranus

Neptune and Pluto

Exploring the universe

Astronautics

Alphabetical index

## THE GHOST PLANET

The existence of **asteroids** was not discovered until the nineteenth century, but astronomers had calculated that there had to be a planet in that area, even though no one had succeeded in finding it. At the end of the eighteenth century several teams of scientists organized a search for the ghost planet. On the first of January, 1801, the Italian astronomer G. Piazzi was studying some stars when he located a luminous body that had not previously appeared on any map, and whose trajectory was different from the trajectories of the stars. That is how he discovered the first **asteroid**. It was given the name **Ceres**, in honor of the goddess of fertility.

Some asteroids don't follow the general orbit of the belt and are known as **Trojans**; they follow the same orbit as Jupiter.

In 1995 the **Galileo** space probe took close-up photos of the asteroids **Gaspar** and **Ida**.

**Ceres** is 488 miles (800 km) in diameter; **Palas,** 275 miles (450 km); **Vesta,** 232 miles (380 km); **Juno,** 116 miles (190 km).

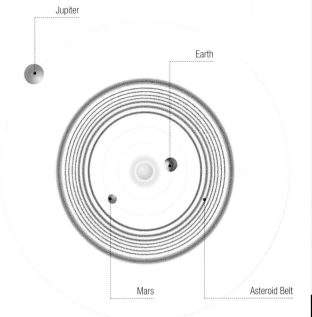

Jupiter

Earth

Mars

Asteroid Belt

Location of the Asteroid Belt. The total mass of all the asteroids doesn't amount to a thousandth of the Earth's mass.

## ORIGIN OF THE ASTEROIDS

According to the most current data, it's now believed that the asteroids are fragments of the initial matter of the solar system; given their insubstantial total mass, they never succeeded in forming a **planet**. Because of the **gravitational pull** of the Sun, these fragments assembled into a common trajectory. It is believed that there are around a million fragments of more than a kilometer in diameter, and a greater number of even smaller pieces. Another hypothesis holds that the asteroids may be left from a planet that was destroyed, but that doesn't seem very likely.

The asteroid that has come closest to the Earth is **Hermes**, a Trojan that came within 475,800 miles (780,000 km) in 1937.

# JUPITER

Jupiter is the largest planet in the solar system; it consists mainly of large masses of **hydrogen** and **helium** in gaseous and liquid forms. It shows characteristic bands formed by the clouds that cover the planet. It has a **ring** around it and a total of sixteen known **satellites**.

## GENERAL CHARACTERISTICS

This giant planet has a very eccentric **orbit**; as a result, sometimes it is relatively close to the Earth, at a distance of only 37,210,000 miles (60 million km). If you observe it with strong binoculars, it is possible to see its **satellites**. Since the planet turns very quickly on its axis, the cloud cover forms characteristic **bands** of dark and light colors. Frequently there are whirlpools in the contact areas between these bands, and they appear as tremendous reddish spots.

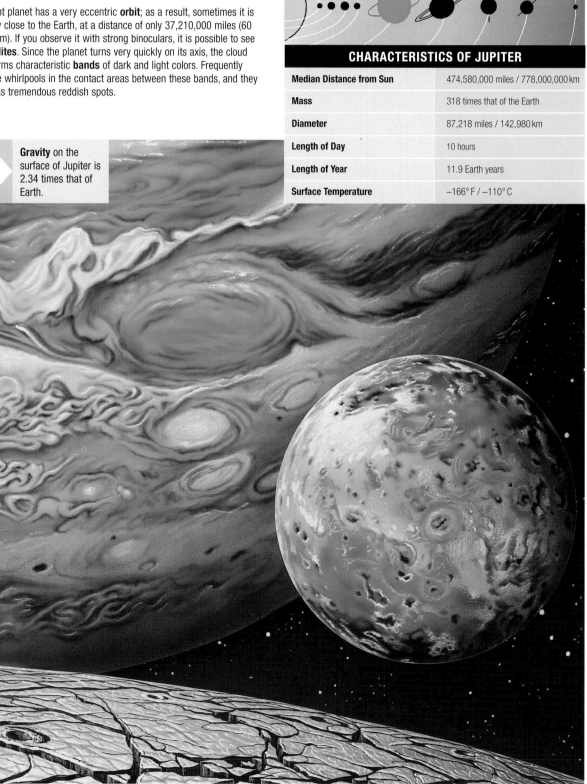

**Gravity** on the surface of Jupiter is 2.34 times that of Earth.

### CHARACTERISTICS OF JUPITER

| | |
|---|---|
| Median Distance from Sun | 474,580,000 miles / 778,000,000 km |
| Mass | 318 times that of the Earth |
| Diameter | 87,218 miles / 142,980 km |
| Length of Day | 10 hours |
| Length of Year | 11.9 Earth years |
| Surface Temperature | −166° F / −110° C |

## COMPOSITION

The interior of the planet consists of a small rocky **core** (1) surrounded by another larger core made up of **metallic hydrogen** (2). Then comes a covering of **liquid hydrogen** (3), and finally a fairly thin layer of gases that make up the **atmosphere** (4). The total composition of Jupiter is 90 percent hydrogen, 5 percent helium, 3 percent methane and ammonia, and the rest of different chemical compounds. The atmosphere forms a dense layer of clouds that make up the characteristic light and dark **bands**.

The rocky core inside the planet makes up just 4 percent of its total mass.

It is believed that the chemical composition of Jupiter is very similar to that of the **original nebula** that gave rise to the solar system.

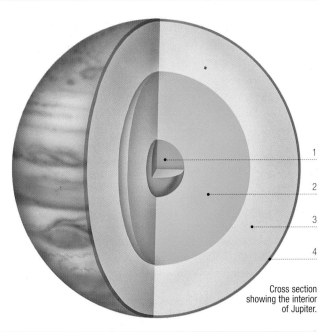

1
2
3
4

Cross section showing the interior of Jupiter.

## SATELLITES

Jupiter is surrounded by a total of sixteen satellites of widely differing dimensions. The largest are **Ganymede** with a diameter of 3,249 miles (5,326 km) (larger than the planet Mercury and the Moon); **Calisto**, with a diameter of 2,928 miles (4,800 km); **Io**, at 2,214 miles (3,630 km); and **Europa**, at 1,915 miles (3,140 km). The smallest is **Lida**, which measures just 10 miles (16 km) in diameter. Io is the satellite closest to the planet, and its shape is irregular because of the planet's gravitational force. Jupiter has many volcanoes and craters caused by meteorites. Europa is covered with a layer of pink ice. Ganymede is also covered with ice, and Calisto is characterized by a great number of small craters that cover its entire surface.

The surface of the satellite Europa is smooth; it is covered by a layer of ice that probably has a layer of liquid water underneath it.

Jupiter gives off radio waves that can be detected with a home radio receiver on the frequency modulation band.

Composite picture showing Jupiter (top left), some of its satellites, and in the foreground, Calisto.

# SATURN

The second-largest planet is one of the most distinctive ones due to the **rings** that surround it. It is also the lightest planet in the solar system. It revolves slowly around the Sun and is made up mainly of **hydrogen** in liquid and gaseous form. It has a dense **atmosphere** of clouds, and eighteen main **satellites**.

## GENERAL CHARACTERISTICS

Slightly smaller than Jupiter, Saturn is characterized by the low **density** of its matter; it is lighter than water and could even float. **Galileo** discovered Saturn's **rings**, but he believed that they were small satellites, as his telescope didn't have enough resolution. Since the planet's rotational axis is tilted twenty-seven degrees, at intervals of several years the rings can be seen almost straight on. Each ring is comprised of small rocks and masses of ice; the largest are only a few feet/meters in length.

Saturn's **gravity** is .93 that of the Earth.

The speed a spaceship needs to escape the gravity of Saturn is 22 miles (35.5 km)/sec.

Saturn with its satellite Dione in the foreground. The width of Saturn's rings is around 43,500 miles (70,000 km), but their thickness scarcely exceeds 12 miles (20 km).

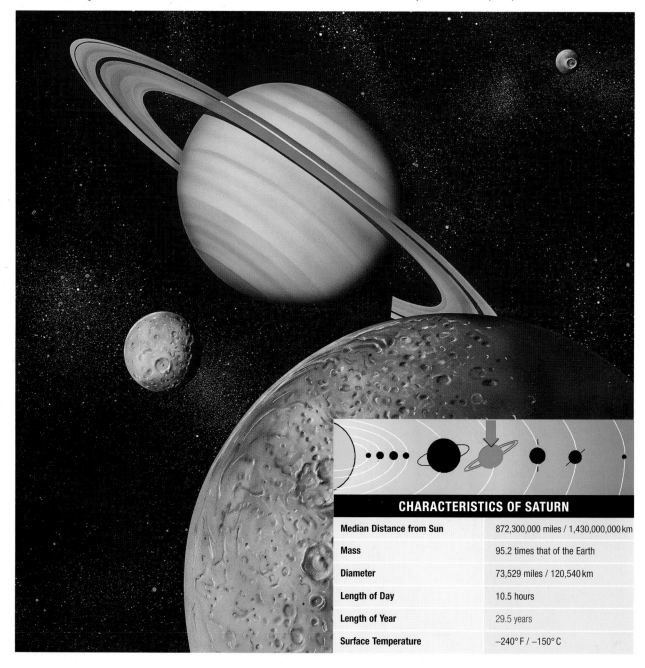

| CHARACTERISTICS OF SATURN | |
|---|---|
| Median Distance from Sun | 872,300,000 miles / 1,430,000,000 km |
| Mass | 95.2 times that of the Earth |
| Diameter | 73,529 miles / 120,540 km |
| Length of Day | 10.5 hours |
| Length of Year | 29.5 years |
| Surface Temperature | −240° F / −150° C |

## COMPOSITION

The planet consists of a small rocky **core** (1) that is smaller than that of Jupiter, followed by a layer of **metallic hydrogen** (2) plus one more layer composed of liquid **molecular hydrogen** (3). Ninety-four percent of the planet is made up of hydrogen, a little more than 5 percent of helium, and the rest of other elements and chemical compounds in very small quantities. There are major storms on the surface, with winds up to 1,100 miles (1,800 km) per hour.

Detail showing the particles of ice and rock that make up the rings of Saturn.

Cross-sectional view showing the interior of the planet Saturn.

1
2
3

It appears that its rings are matter from a satellite that never formed.

## SATELLITES

The planet has eighteen main **satellites,** plus some more smaller ones that are part of the rings. The largest of all is **Titan**, at 3,142 miles (5,150 km) in diameter; it is located at a median distance of 745,420 miles (1,222,000 km) from Saturn. It is characterized by a thick **atmosphere** made up of 97 percent nitrogen, and the rest of methane and other hydrocarbons. It appears that under the dense layer of clouds that covers it there is a huge ocean of methane. In the year 2004 the spaceship *Cassini* will deposit a probe onto this satellite to see what it is like underneath all those clouds.

The space probe *Voyager* has proven that the rings are really a series of many thin rings that are superposed on top of one another.

The satellite **Mimas** measures 239 miles (392 km) in diameter and has a tremendous crater 61 miles (100 km) in diameter, the result of an impact from a meteorite.

# URANUS

Uranus is another of the giant planets, but given its distance from the Earth, it is scarcely visible with the naked eye. It has the same composition as the other **gaseous planets**, and it too rotates rapidly on its axis. It also has several thin **rings** that are invisible through telescopes. It has seventeen **satellites** (including five main ones) that have so far been discovered.

## GENERAL CHARACTERISTICS

Uranus is one of the giant planets, even though it is a bit smaller than Jupiter and Saturn. It was discovered by William **Herschel** in 1781, and through a telescope it appears bluish in color. One of its most notable characteristics is the tilt of its **rotational axis** at 98 degrees, so it practically coincides with the plane of its orbit around the Sun. Its **satellites** are also tilted at the same angle. The precise reason for this peculiar phenomenon is not known; it might be due to some anomaly that occurred when the planet was formed, or the impact from a giant **meteorite** that tipped the planet.

On the surface of Uranus **gravity** is .79 that of the Earth.

The speed a ship needs to reach to escape the gravity of Uranus is 13 miles (21.3 km)/sec.

Uranus, with a bluish color, as seen from its satellite Miranda.

## COMPOSITION

The planet is probably made up of a rocky **core** (1) surrounded by a thick **mantle** (2) of ice, topped off by the gaseous **atmosphere** (3). The components of the atmosphere are the same as for the other giant planets: hydrogen, helium, methane, and other hydrocarbons; however, it appears that the amount of **methane** is greater than that of the other planets because of the greenish color that Uranus has.

## THE RINGS OF URANUS

Astronomers have not succeeded in viewing the **rings** of the planet, but they must exist because of some anomalies visible in the light at certain times when the planet is observed in front of a star. In 1986 the spaceship *Voyager 2* came close enough to photograph the rings. They are a series of ten thin rings made up of small fragments of rock and other very dark matter.

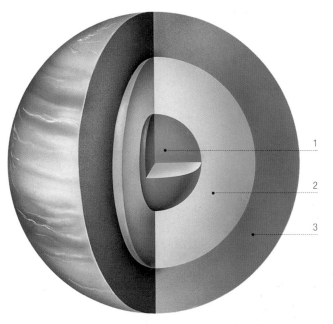

Cross-sectional view showing the interior of the planet Uranus.

Position of the rings of Uranus as determined by the *Voyager* probe.

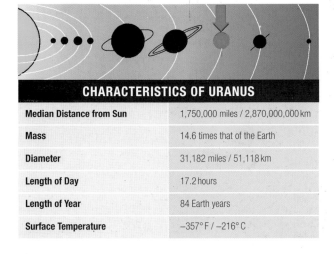

### CHARACTERISTICS OF URANUS

| | |
|---|---|
| **Median Distance from Sun** | 1,750,000 miles / 2,870,000,000 km |
| **Mass** | 14.6 times that of the Earth |
| **Diameter** | 31,182 miles / 51,118 km |
| **Length of Day** | 17.2 hours |
| **Length of Year** | 84 Earth years |
| **Surface Temperature** | −357° F / −216° C |

## SATELLITES

So far five main satellites and twelve smaller ones have been discovered. They are located at a distance between 30,500 and 7,320,000 miles (50,000 and 12,000,000 km). *Voyager* succeeded in photographing the surface of Uranus, which is quite dark and covered with a layer of dark ice. There are also some craters and cracks. **Miranda**, the smallest of the main satellites, measures 287 miles (470 km) in diameter and appears to be divided into several pieces that have fused together.

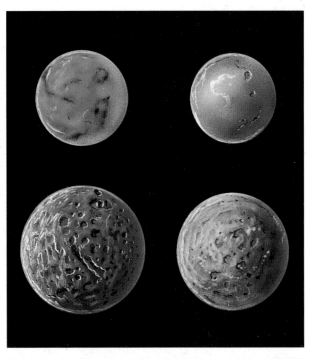

The four largest satellites of Uranus, from left to right and top to bottom: **Ariel** (708 miles / 1,160 km), **Umbriel** (714 miles / 1,170 km), **Titania** (964 miles / 1,580 km), and **Oberon** (927 miles / 1,520 km).

# NEPTUNE

Neptune is a blue **gaseous** planet that can be seen only with the aid of powerful binoculars. It has a very active **atmosphere**, as indicated by its spots and transverse bands. It is surrounded by a small **ring** and has eight **satellites**.

## GENERAL CHARACTERISTICS

The planet consists of a rocky **core** surrounded by a thick **mantle** of ice, covered by the gaseous **atmosphere**. Its composition is similar to that of the other giant planets, consisting mainly of **hydrogen**, **helium**, and **methane**. The **rings** are complete, even though only fragments are visible from Earth. Only two of the planet's eight **satellites** are visible from Earth (**Triton** and **Nereida**); the rest of the satellites were discovered by the spaceship *Voyager 2* in 1989.

On the surface of Neptune **gravity** is 1.12 times that of the Earth.

The speed needed to escape the gravity of Neptune is 14 miles (23.3 km)/sec.

Neptune (in blue) with Triton and Nereida, its main satellites.

**Triton** measures 1,647 miles (2,700 km) in diameter and rotates in a direction opposite that of Neptune; as a result, its speed is gradually diminishing, and within a hundred million years it will fall into the planet and disappear.

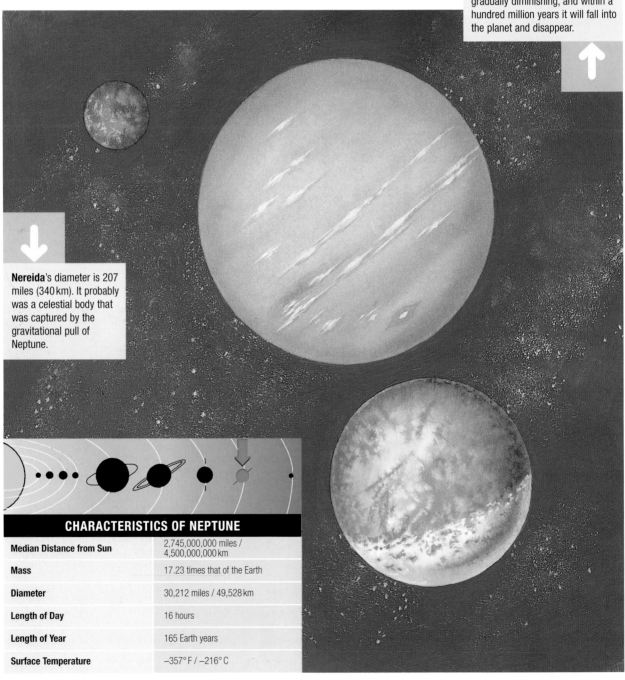

**Nereida**'s diameter is 207 miles (340 km). It probably was a celestial body that was captured by the gravitational pull of Neptune.

| CHARACTERISTICS OF NEPTUNE | |
|---|---|
| **Median Distance from Sun** | 2,745,000,000 miles / 4,500,000,000 km |
| **Mass** | 17.23 times that of the Earth |
| **Diameter** | 30,212 miles / 49,528 km |
| **Length of Day** | 16 hours |
| **Length of Year** | 165 Earth years |
| **Surface Temperature** | −357° F / −216° C |

# PLUTO

Pluto is the most distant of the known planets in the **solar system**, and the existing data on Pluto is still quite scant. In addition, no **spaceship** has ever come close to it. It is a small, rocky planet that evidently has a thin **atmosphere** and one satellite.

## GENERAL CHARACTERISTICS

The planet was discovered in 1930 when some anomalies in the orbit of **Neptune** were identified. Because this planet is so different, it is thought that it could be a wandering planet that got caught by the **gravitational pull** of the **solar system**. It consists of a rocky **core** surrounded by a layer of ice made up of **methane**. The solid surface is also made up of methane, just like the atmosphere. There are dark spots on the surface, and it's probably covered with craters.

On the surface of Pluto gravity is .04 that of the Earth.

The speed a spaceship needs to attain to escape the surface of Pluto is 6.3 miles (10.4 km)/sec.

Pluto, on the right, and its satellite Charon, which revolves around Pluto and always shows the same side.

The satellite **Charon** is 14,640 miles (24,000 km) from the surface of Pluto; it measures 7,320 miles (1,200 km) in diameter, which is exceptionally large for a satellite. It was discovered in 1978.

### CHARACTERISTICS OF PLUTO

| | |
|---|---|
| **Median Distance from Sun** | 3,599,000,000 miles / 5,900,000,000 km |
| **Mass** | .002 times that of the Earth |
| **Diameter** | 1,403 miles / 2,300 km |
| **Length of Day** | 6.4 Earth days |
| **Length of Year** | 248 Earth years |
| **Surface Temperature** | −369° F / −223° C |

# GREAT ASTRONOMERS OF ANCIENT TIMES

**Astronomy** did not exist as such in ancient times. The priests were the ones whose duty it was to observe the sky; they searched in the stars for answers to their daily questions. Still, the great Greek thinkers, the founders of **science,** were the first ones to study the sky and establish the bases of astronomy.

## ARISTOTLE

This Greek philosopher was born in Stagira in 384 B.C. and died in Calcedonia in 322 B.C. He was one of the main thinkers of ancient times and was dedicated to botany, zoology, psychology, medicine, physics, and astronomy, as well as philosophy. Even though the sciences were just one area to which he didn't devote many of his works, his authority as a thinker was so great that for centuries his conclusions in these fields were considered irrefutable.

Aristotle maintained that the **Earth** was a sphere that remained in a fixed position in space and was the center of the **universe**. The other **planets,** the **stars,** the **Moon,** and the **Sun** revolved around it. He proved that through philosophical reasoning, and since no one had access to modern mathematical knowledge or instruments of observation, it was impossible to contradict him.

The things that Aristotle said, in other fields as well as astronomy, were considered absolute truth for almost a thousand years.

Aristotle wrote *About the Sky*, where he addressed topics in astronomy.

In 335 B.C. Aristotle founded the Lyceum in the city of Athens.

Aristotle was a disciple of Plato and a contemporary of Alexander the Great.

## HIPPARCHUS OF NICAE

This Greek scientist lived during the second century B.C.; we don't know many details about his life, but he is considered the founder of **scientific astronomy**. He did some important calculations on the movement of the Sun and the Moon, and described the Moon's **orbit** with some precision.

One of his main works was the first **catalog of stars**. He succeeded in identifying a little more than a thousand stars; he classified them into six categories according to their brightness, and he invented the method that is still in use today.

### STAR CATALOG

Star catalogs are very important in helping us understand how stars change position through time.

# ERATOSTHENES

The Greek geographer, philosopher, and astronomer Eratosthenes was born in Cyrene in 275 B.C. and died in Alexandria in 194 B.C. He was a clever mathematician who performed many very precise calculations to determine geographic distances. Among other things, he invented a type of grid with perpendicular axes that could be used to locate towns and cities when their distances were known. He was director of the **library of Alexandria**.

One of his most important accomplishments was calculating the **circumference of the Earth**. He observed that on the spring **equinox** (March 21) the Sun was reflected on the bottom of the wells in the city of Aswan, but in Alexandria (which was located on the same meridian but a little farther south) there was a little shadow. He deduced that that was due to the **curvature of the Earth**. Then he measured the distance between the two cities and determined the radius of the Earth very accurately.

## THE OBSERVATIONS OF ERATOSTHENES

Even though the Egyptians were very knowledgeable in many sciences, they were not known for their astronomical calculations.

Sun

Aswan well

Obelisk at Alexandria

Shadow

According to the calculations of Eratosthenes, the radius of the Earth was 3,904 miles (6,400 km). This figure is very close to the measurements we make today.

# CLAUDIUS PTOLEMAEUS (PTOLEMY)

This Greek philosopher, mathematician, and astronomer was born and lived in Alexandria in the second century B.C. He wrote a monumental work in thirteen volumes entitled *Sintaxis Matematica*, in which he assembled all the knowledge about **astronomy** that existed at that time. In addition, he created **astronomical tables** and an important work of **cartography** that was used to make the most accurate maps of his time. He also made a **catalog** that includes 1,200 **stars**.

His main contribution to this science is the **planetary model** he created and described in five books. His concept of the universe was adopted by astronomers and lasted for more than thirteen centuries.

## PTOLEMY'S MODEL OF THE UNIVERSE

The system proposed by Ptolemy considered that the **Earth** was the center of the **universe**. He maintained that the Earth was a sphere and that the Moon, the planets, the Sun, and the stars were arranged around it, revolving in precise orbits. In order to explain observed irregularities, Ptolemy invented a complicated set of corrective calculations.

Ptolemy

# MODERN ASTRONOMERS

The ideas about the **universe** that had been conceived by the classical Greek astronomers were considered irrefutable truths for several centuries. The fifteenth century produced the first great revolution when Copernicus stated that the Earth was not the center of the universe. From that time on astronomy began its transformation into the science that we know today.

## NICHOLAS COPERNICUS

A Polish astronomer, Copernicus was born in Torun, on the banks of the Vistula on February 19, 1473, and died in Frombrock on May 24, 1543. He studied law, astronomy, and languages at the universities of Krakow, Bologna, and Padua. Starting in 1512 he was the canon of Frombrock, where he carried out his duties and conducted astronomical observations. He also devised some very useful inventions for the city, such as a hydraulic water supply system.

He was a careful analyst of all the theories known at that time, and he compared them with the most recent available information, and with his own observations. All this led him to conclude that the **Earth** was not the center of the **universe**.

Copernicus assembled his theory in a work that was opposed by the Church, but he saw it published during his lifetime.

Copernicus maintained that the **Earth** and the other **planets** revolved around the Sun.

## JOHANNES KEPLER

This German astronomer was born in Weil der Stadt on December 27, 1571, and died in Ratisbona on November 15, 1630. He worked as the imperial mathematician, but he always had great financial difficulties. He invented a **telescope** in order to carry out his observations more effectively, but he concentrated his work on mathematical calculations for the **trajectories of the planets**; this allowed him to devise his laws concerning their movement.

**Kepler's laws** confirm that the planets move in elliptical orbits around the **Sun**, and the closer they are to it, the faster they move.

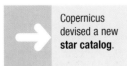
Copernicus devised a new **star catalog**.

Kepler's laws showed that the theories of **Copernicus** were correct.

To come up with his laws, Kepler spent ten years studying the orbit of **Mars**.

## GALILEO GALILEI

This Italian mathematician, astronomer, and physicist was born in Pisa on February 15, 1564, and died near Florence on January 8, 1642. He discovered the laws of the **pendulum**, constructed a hydrostatic balance, and invented a gas thermometer. In 1609 he constructed an improved **telescope** that magnified thirty times and that he used to study the stars.

He made some very important contributions to astronomy, such as the discovery of **sunspots**, calculating the **rotational period** of the Sun, and determining that the stars are very far from our planet, and that the **universe** may be infinite.

He was a great defender of the theory of **Copernicus**, and that led to a confrontation with the Church, which had declared that the Copernican ideas were heresy for denying that the **Earth** was the **center of the universe**. The Inquisition denounced him and put him on trial, and in response to a threat of being sent to prison, in 1632 he was forced to renounce those theories; he was confined to house arrest, where he continued working with his disciples despite being half blind.

Galileo demonstrated that the **Milky Way** is not a cloud but a large number of stars.

According to legend, after renouncing the Copernican theory in front of the judges of the Inquisition, Galileo asserted that "in spite of that (the Earth) does move."

Other discoveries of Galileo include the mountains on the **Moon** and four of **Jupiter's satellites**.

## ANALYZING A STAR

In 1814 the German physicist Joseph **Fraunhofer** observed that as the light from the Sun (or any other star) passed through a prism, black lines were produced (**the spectrum of absorption**), and that makes it possible to analyze the composition of any **star**.

In 1927 Georges **Lemaître** proposed a theory on the origin of the universe that later became known as the **Big Bang.**

Today's astronomers are able to study hundreds of millions of stellar systems located outside our galaxy.

# TELESCOPES AND OTHER INSTRUMENTS

Advances in astronomy are closely linked to the development of scientific instruments that make it possible to study the skies. Still, amateurs can carry out observations using very simple instruments. The instruments available for studying stars run the gamut from **binoculars** up to the large **telescopes** used in astronomical observatories.

## A SIMPLE WAY TO COUNT STARS

All the stars that we can see with the naked eye belong to our galaxy, the **Milky Way**. There is a very simple way to calculate about how many we can see on any night. Cut out a $4^3/_4$-inch (12 cm) circle on a piece of cardboard and hold it 12 inches (30 cm) in front of your eyes; for that purpose you can tie on a 12-inch (30 cm) string and keep it stretched tight between your face and the cardboard.

The hole in the cardboard makes it possible to study 1 percent of the sky. You can count the stars you see through the hole and repeat the count ten times in different directions. By adding together the numbers thus obtained, you will have the number of stars in 10 percent of the celestial sphere. By multiplying by ten you will have calculated the number of stars.

To observe the sky effectively, you have to look for a place far from town to avoid reflections from the lights.

Counting the stars in the sky.

Good binoculars can help you see some surprising details on the Moon.

## BINOCULARS

This optical instrument used for observing wild animals and looking at distant people and places can also be used for studying a fairly large number of stars and the surface of the Moon. With strong binoculars you can see a large number of craters, mountains, and "seas" on our satellite. The best type for astronomical observations is a $7 \times 50$ (seven power with an objective lens of 50 mm diameter).

### NOT AT THE SUN!

**Binoculars** must **never be used** to look at the **Sun**. The result might be irreversible **blindness**.

When you look at stars through binoculars, it is a good idea to rest your elbows on a firm support to reduce vibrations.

# THE ASTRONOMICAL TELESCOPE

This simple instrument, as it was used in ancient times, consists of a tube fitted with a magnifying lens in one end. Later it was discovered that the power could be increased by adding a second lens. Galileo perfected this arrangement and constructed a thirty-power telescope.

The first lens provides a small image of the object under observation—the Moon, for example. The second lens, which is located in the eyepiece (through which the observer peers) magnifies that first image.

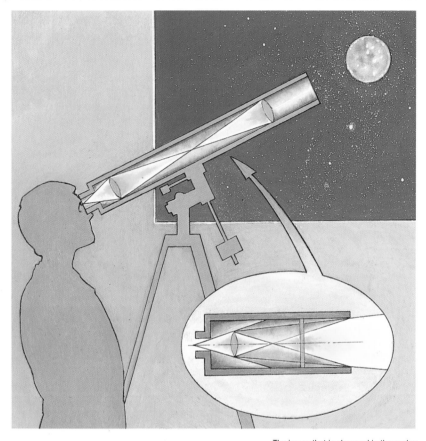

## THE TELESCOPE

This instrument is an improvement on the earlier astronomical telescope; it contains more lenses that correct the defects produced in the image and increase the instrument's capacity. The two most important characteristics of telescopes are their **power** and their **luminosity**. The former is a function of the relationship between the **focal distance** between the objective and ocular lenses. Luminosity depends on the diameter of the **objective lens**; the larger it is, the better the luminosity of the instrument.

The image that is observed in the ocular lens of the astronomical lens appears upside down.

## MODERN TELESCOPES

Modern telescopes are large in size, and consist of many auxiliary features that allow them to be moved with great precision. In addition, instead of observing the image directly, the image is produced by **reflection** in a mirror. The light that enters the telescope is reflected in a **concave mirror** located at the bottom. The reflected image falls onto a flat mirror, and that is what is observed with the aid of the ocular lens.

### RESOLUTION

Resolution is the instrument's capacity to differentiate between two objects placed very close together.

The human eye's power of resolution is referred to as visual acuity.

A telescope with a 1,000 mm focal length objective and a 10 mm ocular focal length yields a magnification of 100 power.

# RADIO TELESCOPES AND SPECTROMETERS

Visible light makes it possible to see a great number of heavenly bodies, but there are many more that emit no light; they seem to be covered by **interstellar dust** that absorbs light, or they are so far away that the power of an **optical telescope** is not sufficient to detect them. As a result, astronomers use other types of **electromagnetic radiation** similar to radio waves. In addition, **spectrometers** make it possible to study the composition of heavenly bodies.

## RADIO TELESCOPES

In 1931 an engineer who was trying to find a way to improve radio reception and avoid static accidentally discovered that some static comes from space. Astronomers began to use this new medium to explore the sky. The result has been the discovery of many stars and other features of the universe.

**Light** is just one part of the **radiation** that celestial bodies give off. The **wavelength** varies along a broad spectrum. Radio telescopes detect all the wavelengths that are not part of visible light. These telescopes are very large and consist mainly of a **reflector** in the shape of a concave mirror to concentrate the radiation in a central point where it is registered—in other words, the **antenna**. From there the signal is sent to an **amplifier** that treats it in such a way that it can be studied.

In order to produce a **resolution** of 1′ a parabolic reflector 2,277 feet (690 m) in diameter is needed.

The Arecibo (Puerto Rico) radio telescope has a parabolic antenna 990 feet (300 m) in diameter.

Radio telescopes are similar to the **parabolic antennas** that we use to view television channels transmitted by satellite, but they are much larger.

# SPECTROMETERS

When white **solar light** is made to pass through a glass prism, the light comes out the other side split into various colors. What happens is that the prism divides the beam of white light into each of its components or colors. Each color has a different **wavelength** and is reflected on the faces of the prism at different angles. All these colors together constitutes what is known as the **spectrum** of solar light.

Physicists have discovered that when a **chemical element** becomes incandescent it gives off a characteristic spectrum—in other words, one that is made up of different proportions of colors. As a result, if we know the spectrum of each element we can determine whether or not the light that comes in from any given location contains that element.

This is the principle of **spectrometry**. A spectrometer is an instrument that analyzes the light that comes in from space (for example, from a star) and makes it possible to know what chemical elements it contains.

The prism decomposes white light into its various constituent colors.

The **wavelengths** of the colors of solar light vary between .40 thousandths of a mm for purple and .70 thousandths of a mm for red.

# INTERFEROMETERS AND RADAR

These two instruments are also used for studying the skies. **Interferometers** consist of two mirrors located at a certain distance from one another that reflect the image or the received radiation onto an **optical telescope** or a **radio telescope** to produce an interference image that improves its **resolution** capacity.

**Radar** sends out a bundle of signals toward a celestial body (the Moon or a planet) and receives an echo that makes it possible to study the surface of that body and determine how far away it is.

Many of these instruments are mounted on **artificial satellites** and **spaceships**, thereby increasing their efficiency by escaping the influence of the layers of air in the Earth's atmosphere.

**Quasars** are powerful emitters of radiation that can be studied with the aid of radio telescopes and spectrometers.

# THE HISTORY OF ASTRONOMY

**Astronomy** was long connected to **astrology**, which is devoted to interpreting human destiny by the stars. The first astronomers were priests who worked in the temples and devised predictions about the future, and also created the bases of this science. The Greeks were the first ones who entirely separated astronomy and religion. That is when the history of astronomical science began.

## BABYLONIA

As early as 5,000 years ago the inhabitants of Babylon (modern-day Iraq), recorded on their tables the regularity of certain celestial phenomena, such as the changes in the **phases of the Moon** and the movement of the **Sun** to different heights at different times of the year. In efforts to predict the future in such practical matters as the flooding of rivers, they would consult the stars, and in order to locate them they devised the first precise measurements of their passage through the heavens.

The Chaldeans invented the water clock to measure time in their observations.

## INDIA

Around the same time when the Chinese were building the first astronomical observatories, Indian scientists were studying **mathematics**, which they related to the stars. **Astronomy** grew out of the old religion of the country, but the necessity of knowing the position of the heavenly bodies soon linked it to mathematics.

Indian mathematicians created the concept of **zero**, a concept that took many centuries to take root in the human mind.

## CHINA

Three thousand years ago the Chinese constructed **astronomical observatories** that were very sophisticated for the time. They divided the year into four seasons that were marked off by the **solstices** (in summer and winter) and the **equinoxes** (in the spring and fall). They divided the heavens into twenty-eight areas that were related to the various positions of the **Moon**. As early as the middle of the eleventh century they observed a nova, the source of the Cancer nebula.

The Indian astronomical observatory of Jantar Mantar in Jaipur, India.

## ASTRONOMERS OF PRE-COLUMBIAN AMERICA

The Mayan pyramid El Adivino in Uxmal, Mexico.

The **Mayas** were great mathematicians who used their knowledge to create a complex and very precise **calendar**. It was very important in predicting the times of harvest and rains. In Mayan cities astronomical observatories were erected next to temples dedicated to the gods; the priests used the observatories to study the movement of the heavenly bodies and make their predictions of eclipses, the movement of the stars, and the length of the year.

The **Aztecs** and the **Incas** also had very precise numerical systems, which constituted one of the bases of their social systems.

## THE GREEKS

The Greeks were excellent mathematicians who applied their knowledge to geometry and astronomy. Much of their knowledge is still used as the basis of science today. **Thales of Miletus**, **Pythagoras**, and **Aristotle** were among the thinkers who established that foundation. The Greek astrologers were able to calculate the **radius of the Earth**, which they already perceived as a sphere. They also calculated very precisely the periods of the planets and many stars.

**Ptolemy** believed that the Earth was the center of the universe and that the Sun and the planets revolved around it.

Around 2,500 years ago the Greeks were the first to completely separate science from religion, thereby freeing up thought and making scientific advances possible.

## COPERNICUS, KEPLER, AND GALILEO

During the sixteenth century these three astronomers did basic work that completely transformed the concept of the universe; building on the knowledge of the Greeks, they provided the definitive impulse to astronomy as a science. After many calculations and observations, **Copernicus** declared that the **Earth** was not the center of the universe, but rather that it and the rest of the planets move around the **Sun**. **Kepler**, with his laws of physics, demonstrated that Copernicus was right. **Galileo** likewise defended his ideas, which he also backed up with his observations; he conducted them with the help of the **telescope** that he invented.

Galileo Gallilei studying the skies.

## MODERN ASTRONOMY

Ever since the time of Copernicus, astronomy has been evolving rapidly. His discoveries tell us that we are merely one planet belonging to a small star located on the outer edge of one of the thousands of **galaxies** that make up the universe. Enormous **optical telescopes** have been constructed to prove that, and they make it possible to view stars located millions of light-years away from the Earth. **Radio telescopes** were invented to see farther, beyond the limits of optical instruments, and radio telescopes have been invented in order to observe the heavens clearly without interference from the Earth's atmosphere. In recent years **artificial satellites** have been launched, and **space stations** have been constructed and equipped with astronomical instruments. Spaceships now travel throughout the entire **solar system**, and they provide us with a unique view of the universe.

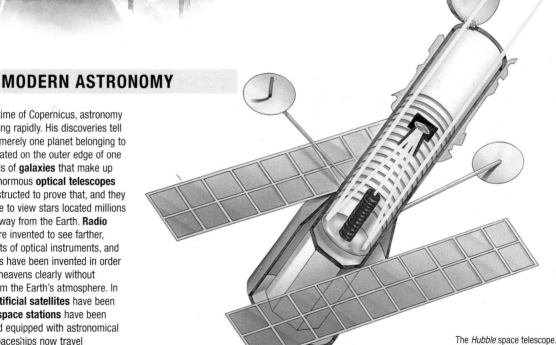
The *Hubble* space telescope.

# THE FIRST ATTEMPTS

In ancient times, humans were inseparably linked to the soil, but they still dreamed of flying. The sky was seen as a sphere that extended over the Earth and where the gods lived, but people still dreamed about going there. The dreams became a reality beginning in the nineteenth century. It became possible to fly, and in the middle of the twentieth century, people succeeded in launching objects outside the planet. Those were the first attempts to make that old dream come true.

## CHINESE ROCKETS

The Chinese invented gunpowder, and they quickly began to search for applications for this extraordinary product. The main uses to which they put it were military, since the possibility of destroying walls and launching projectiles, rudimentary though it was, is an important basis for any army. A hollow tube (such as a piece of bamboo) filled with powder and fitted with an opening in the rear, was propelled forward when the explosive was lit. By placing such a tube on end, it was possible to make it travel upward in a fairly straight line. That's how the first rocket came to be.

The fins of a rocket serve to stabilize it in flight by overcoming wobble and assuring a straight trajectory.

Fireworks are an ancient application of gunpowder. They involve launching a small rocket into the air so that it explodes at a certain altitude and sends out colored sparks in all directions.

## JULES VERNE

This French writer was born in 1828 and died in 1905. He was the author of many adventure books such as *Around the World in Eighty Days, A Journey to the Center of the Earth,* and *Twenty Thousand Leagues Under the Sea.* In all his works he shows a great vision of the future accomplishments of science and technology. In one of these works he delves into space travel a full century before it became possible, performing many very precise calculations. That was in his book *From the Earth to the Moon,* in which a group of people are placed inside an enormous cannon ball and shot toward our satellite.

A scene from the movie *From the Earth to the Moon,* based on the work by Jules Verne.

### TO LEAVE THE EARTH

Among many other calculations, Verne determined that the minimum speed required to escape the Earth's gravity is 6.83 miles (11.2 km) per second.

Jules Verne described the effects of weightlessness on space travelers.

# THE FATHER OF ASTRONAUTICS

The Russian K. E. Tsiolkowski was the first scientist to dedicate himself to the basic problems of space travel. He was born in 1857, and his pioneering studies were ahead of their time; as a result, when he died in 1935 he hadn't seen any of his projects become reality. Among the many subjects that occupied him was the calculation of the right trajectories to escape the Earth's gravity with a minimum expenditure of energy, as well as how much each of the parts of a rocket should weigh in order to blast off. One basic idea he proposed was the use of liquid fuels instead of solid ones.

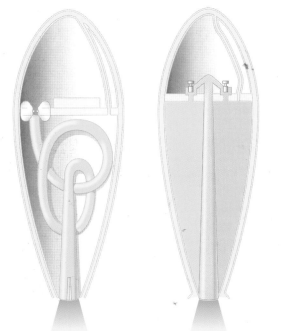

The liquid fuels proposed by Tsiolkowski included hydrogen, oxygen, and carbon hydroxide.

↑

The first rocket prototypes created by Tsiolkowski in 1914 and 1915.

The V-2 flying bomb

Esnault-Pelterie

Goddard

von Braun

Oberth

The deadly V-2 flying bombs were the precursors of present-day missiles; after the war they served as rockets at the start of the Space Race.

## TO THE MOON!

The fantasy of Jules Verne inspired Oberth to create the foundation for a trip to the Moon.

## ESNAULT-PELTERIE AND GODDARD

The Frenchman R. Esnault-Pelterie (1881–1957) concentrated primarily on the problems of aeronautics, but later he focused on astronautics. He designed various propulsion motors and furthered the possibilities of using nuclear fuels.

The American R. H. Goddard (1881–1941) studied numerous problems pertaining to a possible trip to the Moon and succeeded in launching a rocket using liquid oxygen and alcohol fuel, just as Tsiolkowski had proposed years before.

## VON BRAUN AND OBERTH

The German engineer Wernher von Braun (1912–1977) was intensely involved in the design of rockets, and during World War II he constructed the famous V-2. After the war, he moved to the United States, where he participated actively in the space program.

Another German, Hermann Oberth (1894–1979), undertook early studies of the problems inherent in interplanetary travel, after having read the book of Jules Verne in his youth. World War II made it necessary for him to collaborate on the production of the V-2, but after that he continued his studies on spaceship propulsion in the United States.

# THE SPACE RACE

Once World War II was over, the old Allies, the United States and the Soviet Union, competed against one another in many fields as they vied for world supremacy. This struggle extended to all technical and scientific fields, including astronautics. Each of these countries began special programs designed to reach the Moon. That became known as the Space Race.

The gigantic rocket *Saturn 5* (363 feet / 110 m high and 2,700 metric tons in weight) made it possible for humans to reach the Moon.

## HOW DOES A ROCKET WORK?

The first step in the Space Race involved building rockets that were capable of overcoming the Earth's gravity. Goddard had already succeeded in launching a rocket using liquid fuel, and the German V-2s had crossed the English Channel to bomb England. All these rockets, plus the ones that were developed subsequently, work the same way.

A simple rocket consists of a propulsion chamber where the liquid fuel is burned, and that expels gases at more than 5,000 degrees Fahrenheit. For that purpose, the turbines have to be very durable so they don't melt. The escaping gases propel the rocket in the opposite direction. The payload is located at the opposite end; that may be other rockets or a space capsule.

There is no air in space, so the fins do nothing to control the flight. That function is served by small auxiliary motors that accelerate or brake the rocket and make it change direction.

Manned rockets tend to consist of several stages. Each stage is a rocket that pushes the others up to a certain altitude. When its fuel is exhausted the next stage ignites.

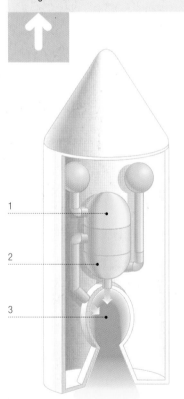

## THE STARTING GUN

The Space Race officially began on July 29, 1955. On that day, the United States announced that it would construct and launch into space an artificial satellite that would enter into orbit around the Earth and take photographs of our planet. That was its contribution to the International Geophysical Year that was to be celebrated between 1957 and 1958, during which scientists from all around the world would carry out measurements and analyses of our planet to improve our knowledge of it.

On August 1, 1955, three days after the American announcement, the Soviets also announced their intention to build and put into orbit a similar satellite.

The principle of the **propulsion rocket**. The oxidizing agent (1) makes it possible for the fuel to burn; the pump or gas impeller (2) pumps the fuel from the tank to the motor (3), where the fuel generates gas as it burns.

## THE FIRST ARTIFICIAL SATELLITE

On October 4, 1957, Soviet scientists put into orbit the first artificial satellite to orbit around the Earth; it was named *Sputnik I*. The beeping sound that it made, which was reproduced on radio and received by radio afficionados, became a sensation all over the world. Space exploration had begun. That first satellite was small, no bigger than a large ball; it was silver in color and had several antennas.

Four months after *Sputnik I* was launched (in the illustration) the first American satellite, *Explorer I*, was put into orbit.

## THE FIRST LIVING CREATURE IN SPACE

On November 3, 1957, the Russians put a new satellite into orbit; it was named *Sputnik II*, and it carried inside it a Siberian Husky named *Laika*, who became the first living being to leave the Earth and venture into space. The dog was fitted with devices that monitored its vital signs and showed the scientists how an animal could survive at those altitudes inside the capsule, and thus, how a human could do it.

### THE DOG, LAIKA, INSIDE *SPUTNIK II*

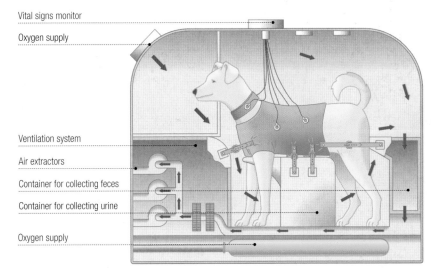

Vital signs monitor
Oxygen supply
Ventilation system
Air extractors
Container for collecting feces
Container for collecting urine
Oxygen supply

## TRAVELING MONKEYS AND RATS

While the Soviet scientists were using Siberian Huskies as astronauts, the Americans opted to use rats and monkeys. In the summer of 1958 Americans launched a rocket that carried the rat *Mia* up to an altitude of 4,880 miles (8,000 km), but the small rocket was lost in the ocean when it returned. Another rat named *Wickie* returned successfully, but didn't provide much information because the monitors weren't working properly. Flights involving monkeys were more successful; they provided valuable information for subsequent manned flights.

NASA (the National Aeronautics and Space Administration) was formed in July, 1958; it is the American civilian organization that deals with all matters involving aeronautics.

The first director of NASA, who served until 1972, was Wernher von Braun.

# MEN IN SPACE

When construction techniques for rockets had reached a certain level that allowed launching a large capsule, and experiments with animals had demonstrated that it was possible to live in space, the two great national powers went a step farther, and for the first time humans succeeded in leaving the planet. A new age had begun.

## YURI GAGARIN

The pilot Yuri Gagarin, who was born in 1934, was one of the first people chosen to make up the Soviet team of astronauts. Because of his technical knowledge, his experience, and his physical condition, he became the first person to travel into space, on April 12, 1961. He did so on board a small capsule named *Vostok 1*. His flight lasted just a few hours, during which he made three revolutions around the Earth; it was enough for him to be acclaimed a great hero upon his return. He was killed in 1968 when the plane he was testing crashed.

Yuri Gagarin, the first man in space

In 1998, at the age of 78, John Glenn returned to space to test the effects of weightlessness on older people.

## JOHN GLENN

John H. Glenn, who was a Navy pilot, was the first American to travel into space. He did so on February 20, 1962, also on board a small capsule in the *Mercury* series. He likewise made a couple of trips around the Earth, and he remained in communication with ground control from his spaceship; these communications were broadcast to the public, and caused a major sensation.

## THE *MERCURY* PROJECT

Many other launches, both Soviet and American, followed the first two manned flights into space. In all these cases, the capsules were tight quarters that had room for only one crew member. All these flights involved testing different systems for the increasingly complex capsules that were to follow. The American flights involving a single crew member were designated the *Mercury* project.

A *Mercury* capsule after landing at sea.

## THE *APOLLO* MISSION

After the success of the first manned space flights, NASA began the *Apollo* project at the end of 1966, with the ambitious plan of putting a man on the Moon. Until that was achieved three years later, a number of increasingly complex tests were conducted by launching capsules containing two or three astronauts, having one of them leave the spacecraft for the first time, and sending satellites to the Moon to explore the terrain before making a manned landing. The last preparatory step involved a manned flight around our satellite.

A Soviet space capsule landing in Siberia.

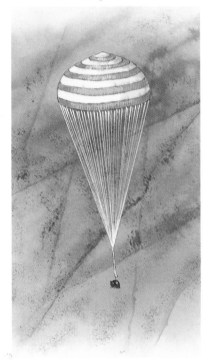

An American space capsule landing in the ocean.

## VICTIMS OF THE SPACE RACE

On January 27, 1967, before the launching of the first *Apollo* spaceship, an equipment failure in the rocket equipment caused a fire, and the three astronauts, Grissom, Chafee, and White, were burned to death. In mid-1967 the Soviet cosmonaut V. Komarov was killed when the parachute of the first *Soyuz* spaceship failed to open and the spacecraft crashed on its return to Earth.

The worst disaster, however, occurred several years later, on January 28, 1987, when the *Challenger* spacecraft exploded in flight seventy-two seconds after takeoff, and its seven crew members were killed.

## PREPARING FOR THE GREAT VOYAGE

In 1963 America instituted the *Gemini* Program, which consisted of launching manned space capsules containing two astronauts. Within a few months, the Soviets did the same.

On July 31, 1964, the American space probe *Ranger 7* crashed onto the surface of the Moon after taking four thousand close-up photographs.

On March 18, 1965, the Soviet cosmonaut Leonov completed the first space walk connected to his spaceship by a cable.

In December, 1965, two American spaceships, *Gemini 6* and *7* docked together for the first time in space, thereby demonstrating the possibility of carrying out this type of operation.

On February 3, 1966, the Soviet space probe *Luna 9* landed gently on the surface of the Moon and began to transmit photographs. In June of the same year, the American probe *Surveyor 1* also landed on the Moon.

# LANDING ON THE MOON

All the experience gained over a little more than two decades was used to achieve one of mankind's mythical dreams: landing on the Moon. The first artificial satellites, starting with *Sputnik I*, had helped open the way. They provided essential data on space and how to survive in it. Manned space flights put the scientists' theories into practice, and once everything was confirmed, there remained only the great leap of transporting a human onto our satellite.

## AIMING FOR THE MOON

After the first non-manned spaceships landed on the lunar surface and sent back data about its composition and topography, the next step was to send up astronauts.

On December 25, 1968, astronauts Borman, Lovell, and Anders arrived close to the Moon in their spaceship *Apollo 8*, entered an orbit, and made several revolutions around it. They succeeded in making direct observations of the surface and taking new photographs so that scientists could look for a suitable landing site.

In March, 1969, *Apollo 9* carried out docking experiments involving a lunar vehicle and the spaceship. The vehicle remained in stationary orbit around the Earth after the flight.

In May, 1969, *Apollo 10* picked up the lunar module and traveled to the Moon. There the spacecraft and the lunar module were uncoupled and the astronauts descended to 4.2 miles (7 km) from the lunar surface; however, they returned to the spacecraft later without having landed on the Moon. That way they confirmed that all the equipment was working properly.

The next step was the definitive one.

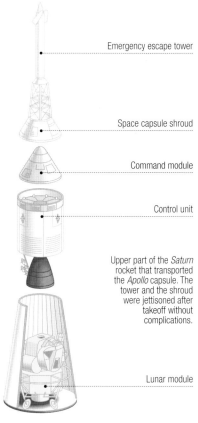

Emergency escape tower

Space capsule shroud

Command module

Control unit

Upper part of the *Saturn* rocket that transported the *Apollo* capsule. The tower and the shroud were jettisoned after takeoff without complications.

Lunar module

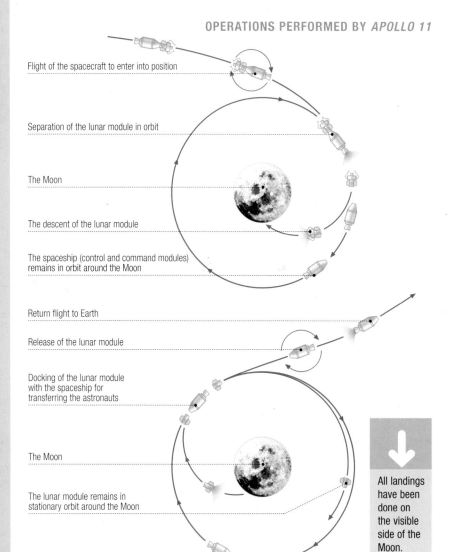

OPERATIONS PERFORMED BY *APOLLO 11*

Flight of the spacecraft to enter into position

Separation of the lunar module in orbit

The Moon

The descent of the lunar module

The spaceship (control and command modules) remains in orbit around the Moon

Return flight to Earth

Release of the lunar module

Docking of the lunar module with the spaceship for transferring the astronauts

The Moon

The lunar module remains in stationary orbit around the Moon

All landings have been done on the visible side of the Moon.

## THE DEFINITIVE FLIGHT

On July 16, 1969, *Apollo 11* blasted off from Cape Kennedy en route to the Moon. Three days later it entered orbit around our planet. The astronaut Michael Collins remained in the *Columbia* capsule to control the operation.

On July 20, Neil Armstrong and Edwin Aldrin descended to the surface of the Moon onboard the lunar module *Eagle* and made a soft landing.

In the early hours of July 21, Armstrong opened the hatch, climbed down the ladder, and stepped onto the surface of the Moon, followed shortly by Aldrin.

They took several walks in the vicinity, scooped up samples of the lunar soil, and set up several instruments they had brought from Earth. When their mission was complete, they climbed back into the lunar module *Eagle*, blasted off, and docked once again with the *Columbia* capsule, returning to Earth without difficulty.

# A HISTORIC STEP

When Neil Armstrong climbed down the ladder of the lunar module and stepped onto the ground, the module's television cameras captured the moment and transmitted it to Earth, where hundreds of millions of spectators were able to follow the event directly. The astronauts were moving around and were able to jump effortlessly due to the reduced gravitational force.

Neil Armstrong said these famous words on television: "That's one small step for [a] man, one giant leap for mankind."

## SIX TIMES ON THE MOON

There have been six other manned expeditions to the Moon (*Apollo 12* to *Apollo 17*) since the first landing.

Many had feared that the surface of the Moon was made of dust and that any spacecraft that landed would immediately sink. Experience showed otherwise.

## A LUNAR VEHICLE

In the most recent trips to the Moon, the astronauts used a vehicle to travel several miles/kilometers from the lunar module.

Between 1970 and 1971 the Soviets sent robots to the Moon in the *Lunik* spaceships, which gathered samples and left scientific instruments behind.

# SPACE EXPLORATION

In the 1960s the conquest of the Moon was one of the most important programs carried out by NASA and by Soviet scientists. But at the same time there were other projects of greater scope and duration that targeted longer-term results: space exploration beyond the Moon.

## CONQUERING THE PLANETS

Between the years of 1962 and 1973 both space powers launched numerous satellites headed for the main planets of our solar system; they obtained closer photographs and some data about their surface. In subsequent years it has finally been possible to use the photos in drawing up maps of those planets, and to touch down on some of them.

 The first satellites from 1957–1958 discovered that the Earth is surrounded by two belts of magnetic fields; these have been named Van Allen's Belts.

In 1997 *Mars Pathfinder* deposited a vehicle onto the surface of Mars.

## LIFE ON MARS?

In 1970 the Russian spacecraft *Venera* landed successfully on the surface of Mars, and other spaceships in this series repeated the feat in following years. The American spacecraft *Viking I* and *Viking II* arrived on Mars in 1975 and carried out experiments to search for signs of life, but they found none.

*Voyager I* and *II* were launched in 1977. They passed close to Jupiter In 1979; Saturn in 1980 and 1981; Uranus in 1986; and Neptune in 1989.

## PROBING VENUS

In 1978 *Pioneer Venus* launched several probes on parachutes onto the surface of Venus

The space probe *Viking* approaching Saturn.

Introduction

Space

The solar
system

The Sun

Mercury

Venus

Earth

Mars

Asteroids

Jupiter

Saturn

Uranus

Neptune
and Pluto

Exploring
the universe

**Astronautics**

Alphabetical
index

# INTERNATIONAL COOPERATION

The Space Race was an extremely costly project, and international financial problems caused both the Russians and the Americans to suspend some of their programs. That opened the door for cooperation with other countries, including the European countries that belonged to the ESA (European Space Agency). Ever since the end of the 1970s, various joint space probes and scientific satellites have succeeded in expanding our knowledge of the solar system, with results that are sometimes spectacular.

*Skylab*, the first American space station was put into orbit in 1973; as a result it became possible for teams of astronauts to stay in space for several months.

The first successful European rocket in the *Ariane* series was launched in 1979.

## THE MODERN ULYSSES

The European space probe *Ulysses* succeeded in viewing the south pole of the Sun.

On March 14, 1986, the European space probe *Giotto* met up with Halley's Comet and succeeded in photographing it and analyzing it up close.

The joint American and European *Hubble* space telescope, which was launched in 1990, produced excellent photographs of new galaxies in 1995; in 1996 it proved the existence of planets around other stars.

On November 28, 1983, the American transport *Columbia* put the European *Spacelab* into orbit.

# SPACE LABORATORIES AND STATIONS

Since outer space has no atmospheric layer like the one that covers the Earth, it offers many advantages for astronomical observations. As a result, several laboratories have been constructed and placed into stationary orbit around our planet. The smallest of them are operated from ground control, but the most elaborate ones have been incorporated into space stations; in other words, permanent laboratories occupied by astronauts who stay in them for several months as they carry on their work.

Space launches such as *Discovery* perform regular flights between these stations and the Earth, transporting materials and astronauts. The goal is to construct a large space station where the astronauts can carry on their work under favorable conditions, and that will serve as an intermediate step for more distant expeditions.

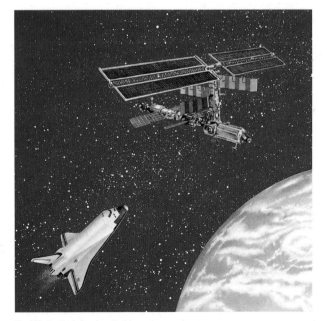

The ISS, the first step in colonizing space.

## AN INTERNATIONAL SPACE STATION

Construction work on the ISS (International Space Station) began in 1998; it involves cooperation among Americans, Russians, and Europeans. At the start of 2001 parts of the ISS space station were ready for habitation.

The first Russian space station was *Salyut*, which was put into orbit shortly after the American station. The Russian space station *Mir* began work in the 1980s as a substitute for *Salyut*, and it remained in use nearly until 2000.

On June 29, 1995, the American space launch *Atlantis* and the Russian space station *Mir* successfully docked at an altitude of 244 miles (400 km) and a speed of over 17,000 miles (28,000 km)/hour; then the astronauts passed from one spacecraft to the other.

# THE FUTURE OF ASTRONAUTICS

Scarcely a dozen years passed from the launching of the first man-made satellite—a simple metal sphere that sent out a beep—to the landing of men on the Moon, and from that time onward space has been filled with satellites, spaceships, and space stations. Proposals even include colonizing the Moon and establishing bases on nearby planets such as Mars.

## PROJECTS UNDERWAY

In the past two decades aeronautics has been characterized chiefly by international cooperation, which has intensified since the collapse of the Soviet Union. The United States (NASA), Russia, Europe (ESA), and Japan (NASDA) are working together on projects that require great economic investments and careful scientific and technical research from hundreds of teams spread around the globe.

The *Cluster II* mission of NASA and ESA was begun in July and August of 2000. This involves four satellites traveling in formation to measure the solar magnetic fields and storms.

At the end of 2001 a geological map of the surface of Mars was completed with the help of *Mars Surveyor*, and in 2003 several probes will land on the red planet in order to gather more samples of the soil and continue studies for installing a possible base.

The *Viking* probe exploring the asteroids while a vehicle studies the surface and composition of a single asteroid.

## SO NEAR, YET SO FAR

While ESA's Integral Astronomical Satellite, which was launched at the end of 2001, records emissions of cosmic rays to verify the origin of the universe, many of the satellites launched throughout 2001 are intended to study Earth: the ozone layer, weather forecasting, studies of glaciers, measuring vegetation, climatic changes, and so forth.

### A SPACE TOURIST

In April, 2001, an American multimillionaire, Dennis Tito, became the first space tourist.

The new instruments for exploring and studying space that are now available guarantee that many of the mysteries of the universe will be revealed in the present decade.

## ROUND TRIP

It is planned that a space probe launched in 2002 will land on the asteroid *Nereus*, gather samples of the soil, and return to Earth in 2008.

A study of Saturn's rings is planned for 2004; a probe will be sent to its satellite *Titan*, where it will land on the surface.

The surface of Mars contains lots of metallic oxides, which give it the red color; they may be a source of oxygen.

## A BASE ON THE MOON

The desire to colonize other heavenly bodies will probably begin with the Moon. The exploration of its surface carried out by the astronauts, the numerous analyses and studies conducted by robots and stationary satellites since that time, and instruments placed on its surface will help make that project a reality. The second planet to be colonized will probably be Mars, which offers similar conditions for constructing bases.

Vestiges of ice were discovered on the surface of the Moon in 1999, so it may contain some water in frozen form.

Space stations around the Earth will be the first step in interplanetary travel; they will probably use the Moon as a launching site, as shown in the illustration.

# ASTRONAUTS

Space exploration is a highly technical activity, the success of which depends heavily on robots and computers. Still, the human presence is required for many activities. Astronauts are the ones to carry out those tasks, and for that they need very special training.

## SPACE SUITS

Space suits have evolved a lot over the years. Special fabrics have been devised, and the helmets have been improved to provide the proper environment for the person who has to work inside the suit. These suits are designed for performing tasks in space outside the spaceship and for moving about the surface of celestial bodies. The first astronauts were attached to their spaceship by cables. Now they have devices that they can control for moving freely around them; but for safety reasons, the cables are still used.

It is necessary to maintain a pressure of around 30 pounds (13.6 kg) per square inch (760 mm) inside both the spaceship and the space suit.

### THE SPACE HELMET

The space helmet has a face shield that protects the astronaut against both the intense visible light given off by the Sun and dangerous cosmic radiation.

### AN ASTRONAUT'S SPACE SUIT

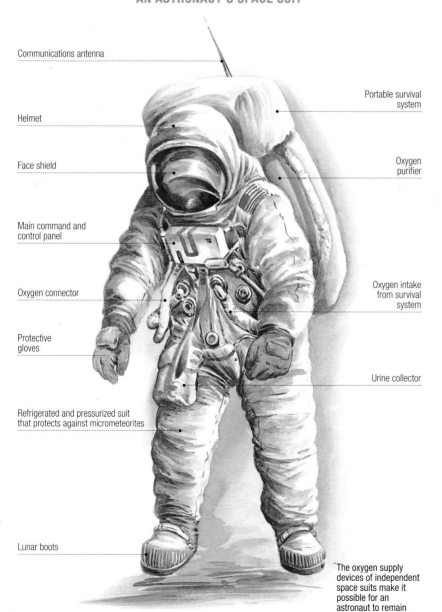

Communications antenna

Helmet

Face shield

Main command and control panel

Oxygen connector

Protective gloves

Refrigerated and pressurized suit that protects against micrometeorites

Lunar boots

Portable survival system

Oxygen purifier

Oxygen intake from survival system

Urine collector

Communication devices are essential for astronauts outside their spacecraft.

Space suits are made from a fine protective fabric that is flexible, lightweight, and very resistant to small rocks that may be floating in space.

The oxygen supply devices of independent space suits make it possible for an astronaut to remain outside the spacecraft for several hours.

# LIFE IN A SPACESHIP

The first space flights lasted just a few hours, but as they grew longer it became necessary to provide more complex equipment to assure that conditions remained acceptable for the astronauts. In the first space stations, such as the Russian *Mir*, the astronauts didn't have much space for moving around freely, and they had to do all their activities in an all-purpose compartment. However, the new space stations such as the International Space Station (ISS) currently under construction are complex installations that are comparable to the ones that exist on Earth in areas of extreme conditions, such as Antarctica. The ISS consists of several modules where the astronauts work, sleep, and have adequate room for rest breaks.

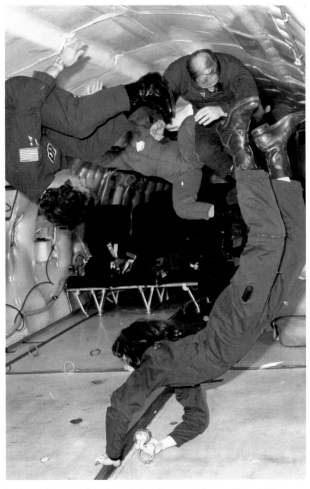

Inside a training room for astronauts; here they are getting used to weightlessness.

During a space mission, each astronaut is responsible for several specific tasks. The greatest difficulty they have seems to be living together for several days in such a restricted area.

## DIFFICULT TRAINING

In order to travel into space, one has to be in very fine physical condition, since the body is subjected to great stress during the launch. In addition, during extended stays in space, the absence of Earth's gravity produces major physiological changes that have to be offset by good training. That is why the first astronauts were test pilots accustomed to experiencing powerful acceleration.

Still, the number of scientists who take part in space missions has been increasing since the 1990s. Although they have to be in perfect health, the new technology facilitates their adaptation to space.

Future travelers who go to the Moon and Mars as tourists will have to be healthy, but the spacecraft will compensate for many of the present inconveniences by creating artificial gravity, among other things.

During the launch of a spaceship, the human body weighs up to ten times more than normal.

Weightlessness, the absence of gravity, causes decalcification of the bones after an extended period of time.

### MUSCULAR WEAKNESS

Astronauts who remain in space for several months suffer from muscular weakness because of the lack of gravity, and when they return to Earth, they have to be carried on a stretcher until they regain their muscle tone.

# ALPHABETICAL INDEX